CONTINUITY AND CHANGE IN THE DEVELOPMENT OF RUSSELL'S PHILOSOPHY

Nijhoff International Philosophy Series

VOLUME 50

The titles published in this series are listed at the end of this volume.

CONTINUITY AND CHANGE IN THE DEVELOPMENT OF RUSSELL'S PHILOSOPHY

PAUL J. HAGER
Faculty of Education, University of Technology Sydney, Australia

Kluwer Academic Publishers
Dordrecht / Boston / London

Library of Congress Cataloging-in-Publication Data

Hager, Paul J.
Continuity and change in the development of Russell's philosophy
by Paul J. Hager.
p. cm. -- (Nijhoff international philosophy series ; v. 50)
Includes bibliographical references and index.
ISBN 0-7923-2688-1 (HB : alk. paper)
1. Russell, Bertrand, 1872-1970. I. Title. II. Series.
B1649.R94H29 1994
192--dc20 93-44873

ISBN 0-7923-2688-1

Published by Kluwer Academic Publishers,
P.O. Box 17, 3300 AA Dordrecht, The Netherlands.

Kluwer Academic Publishers incorporates
the publishing programmes of
D. Reidel, Martinus Nijhoff, Dr W. Junk and MTP Press.

Sold and distributed in the U.S.A. and Canada
by Kluwer Academic Publishers Group,
101 Philip Drive, Norwell, MA 02061, U.S.A.

In all other countries, sold and distributed
by Kluwer Academic Publishers,
P.O. Box 322, 3300 AH Dordrecht, The Netherlands.

Printed on acid-free paper

Printed in the Netherlands

This book is dedicated affectionately to
Maureen
whose unfailing love and support have
sustained me over many years

Table of Contents

"In matters of *work* my life has had very great continuity and unconscious unity"

(Russell in a letter of 8 February 1913 to Ottoline Morrell)

"Where we are inquiring into the opinions of a truly eminent philosopher, it is probable that these options will form, in the main, a closely connected system, and that, by learning to understand them, we shall ourselves acquire knowledge of important philosophic truths."

(Russell, September 1900, Preface to first edition of *A Critical Exposition of the Philosophy of Leibniz*)

Preface

The general view of Russell's work amongst philosophers has been that repeatedly, during his long and distinguished career, crucial changes of mind on fundamental points were significant enough to cause him to successively adopt a diversity of radically new philosophical positions. Thus Russell is seen to have embraced and then abandoned, amongst others, neo-Hegelianism, Platonic realism, phenomenalism and logical atomism, before settling finally on a form of neutral monism that philosophers have generally found to be incredible. This view of Russell is captured in C.D. Broad's famous remark that "Mr. Russell produces a different system of philosophy every few years . . ." (Muirhead, 1924: 79). Reflecting this picture of Russell continually changing his position, books and papers on Russell's philosophy have typically belonged to one of two kinds. Either they have concentrated on particular periods of his thought that are taken to be especially significant, or, accepting the view of his successive conversion to distinctly different philosophical positions, they have provided some account of each of these supposedly disconnected periods of his thought. While much good work has been done on Russell's philosophy, this framework has had its limitations, the main one being that it conceals the basic continuity behind his thought. Examples of the first kind of book or paper on Russell's philosophy abound, with Griffin's *Russell's Idealist Apprenticeship* being particularly notable; examples of the more comprehensive second kind, that aspire to do justice to all periods, are somewhat fewer, with Jager's *The Development of Bertrand Russell's Philosophy* being noteworthy. A further legacy of the wide acceptance of the view that Russell successively espoused distinctly different philosophical positions, is that particular works have been canonised in some quarters as the 'real' Russell, to the detriment of his other works. "On Denoting", *Principia Mathematica* and the works on logical atomism have been favourites in this respect. Nevertheless, in recent years there have been signs of increasing awareness that the extent to which Russell was prone to change his philosophical position may have been greatly exaggerated.

This book represents the first detailed attempt to trace the fundamental unity that lies within all of Russell's philosophical work, as well as the reasons behind those limited orderly changes that did, in fact, occur within it. As such the book is addressed primarily to serious students of Russell's philosophy, and is suitable for use in both postgraduate and undergraduate courses on Russell's philosophy generally, on specific aspects of his work, or on analytic philosophy in the twentieth century. However, it is the author's hope that the book is written clearly enough to be approachable by the many general readers likely to be interested in finding out more about Russell's philosophy.

Thus, the main thesis of the book is that there is a lot more *continuity* in Russell's philosophy than has usually been acknowledged, and that the major *changes* that do occur are much more orderly than Russell's reputation for erratically changing his views allows. In Part One it is maintained that the continuity of his philosophy centres on the method of *analysis* and that this method of analysis has not been well enough explicated by previous commentators. Drawing on a wide selection of Russell's own statements, a general account of Russellian analysis is developed which shows it to have a highly organised structure, which he consistently applies throughout all of his post-idealist philosophising. One outcome of this is that some significant misunderstandings of Russell's philosophy are traced to simplistic notions about analysis. Another outcome of recognising the detailed structure of Russellian analysis is that the vital role of *relations* in his philosophy is highlighted. It turns out that in certain logical constructions and series, relations have a significance that has been generally overlooked, thereby generating further widespread misunderstanding of Russell's philosophy. However, while the method of analysis underlines the prominence of relations in Russell's philosophy, the *sorts* of relations that are most important change over time. Accordingly, Part Two seeks to identify the basis of these changes in his philosophy. It is found that there are historical reasons, largely to do with his realism being a reaction to Kant, why *space and time* theories are absolutely central to Russell's philosophy. In 1914, and again in 1919, he was faced with totally compelling reasons to alter his views on space and time. In both cases the change was drastic enough to affect most areas of his philosophy, thereby explaining why different sorts of relations figure prominently in different phases of Russell's philosophical development.

This work grew out of my PhD studies at the University of Sydney. Special thanks are due to Professors David Armstrong and Keith Campbell for their encouragement and capable supervision. I also received significant help from the Bertrand Russell Archives at McMaster University, especially from Dr. Ken Blackwell, the Russell Archivist, and Professor Nick Griffin. I am grateful to the University of Technology, Sydney for granting me study leave for autumn semester 1993, thereby enabling me to visit the Bertrand Russell Archives to complete this book. My thanks also go to Maree Joulian, University of Technology, Sydney, who prepared the diagrams. As well, I am indebted to Jane Hager, Paul M. Hager and Gregory Hager for various help in preparing the manuscript. Finally thanks to the Bertrand Russell Archives and the University of Texas at Austin for permission to quote from Russell's letter of 8 February 1913 to Lady Ottoline Morrell.

Oatley, Sydney
October 1993

P. J. H.

CHAPTER 1

Introduction

It has been a striking fact about the reception of Russell's philosophy that while the works prior to 1920 have been enormously influential and widely discussed, his later works have been comparatively neglected. Russell himself complained bitterly that these later works had been misunderstood.[1] Given that Russell ranks at or near the top of most people's list of the most important twentieth century philosophers, I find this reception of his later work very puzzling. Of course this situation was encouraged by the once widespread belief that Russell was prone to producing new philosophical positions that had little or no connection with his earlier views. Russell himself repeatedly asserted that there was a fundamental continuity underlying the changes and since his death in 1970 this view seems to be gaining some acceptance. This is evidenced by the slowly increasing interest being shown in his later works, which remain relatively unknown in comparison to the works up till about 1920.[2] Further interest in the continuity underlying changes in Russell's philosophy is evident from the many papers and books generated from research into Russell's previously unpublished work held at the Bertrand Russell Archives, McMaster University. This research effort has been particularly concerned with the evolution of Russell's thought from neo-Hegelianism to the realism of *Principles of Mathematics* through to the more austere *Principia Mathematica*. The recent appearance of Russell's previously unpublished papers from this period (in the *Collected Papers*) will likely stimulate further understanding of the continuity that underlies the major changes in this important period of Russell's philosophical work.

So while much has been achieved, a lot of work still remains to be done to identify the basis of the continuity in Russell's overall philosophy. This book aims to be a contribution to that task. It seeks to specify in detail a method of philosophising that underlies all of Russell's mature work. From the stand point of this core of stability, it emerges that the major changes in Russell's views were much more orderly than has usually been supposed. So this book is more concerned with getting Russell right than with criticising him. In this I am in complete agreement with Alan Wood (whose untimely death curtailed what promised to be an outstanding account of the development of Russell's philosophy):

> I have come across few present-day critics of Russell who do not either unwittingly repeat points he himself has made, or else show ignorance of his true views. (It must be remembered that his books are already becoming classics; and a classic may be defined as a book which people think they know about without having read it.)
>
> The need at the moment, therefore, is not to criticise Russell, but to understand him.
>
> (*MPD*: 191).

Another implication of my concern to get Russell right is that there will be extensive quoting from a wide range of his works. This is necessary to justify the claim that Russell did indeed proceed in the way described.

In attempting to spell out the basis of *continuity* in Russell's philosophy, I will focus on the method of *analysis* and the role of *relations* in such analysis. *Change* in his philosophy, on the other hand, will be found to centre on his theories of space and time. Changes in the ontology of space and time turn out to be the main stimulus to other changes in Russell's philosophical theories. 'Analysis' and 'relations' would certainly be very high on any list of key words that describe Russell's seminal contribution to the new direction that British philosophy took at the beginning of this century. By now it is firmly entrenched in philosophical lore that Russell, stimulated by Moore, opposed the holistic, idealist approach to philosophical problems with a piecemeal, realist stance whereby wholes were *analysed* into parts, problems into sub-problems, and broad questions into several more specific ones. Likewise, as against the idealist doctrine that all propositions are of the subject-predicate form, Russell argued that some propositions are irreducibly *relational*, a point of profound philosophical importance.

So much hardly bears repeating, yet as so often with the obvious, light soon shades into dark once some pointed questions are asked. Given the enormous secondary literature on Russell's philosophy, one would expect the topic of *analysis* to be well and truly elaborated. In fact, surprisingly little has been written on the subject, whilst the divergent views of those who have written signal the need for further work in this vital area. If anything, the central role of *relations* in Russell's philosophy has received even less attention than analysis, with the same inconclusive results.

In his classic paper on Russellian analysis, Weitz (1944: 110) claimed that it

> . . . is a curious fact that Russell, one of the greatest modern exponents of the method of analysis, has never discussed in any detail what he means by it.

Attempting to make up for Russell's alleged omission, Weitz (1944: 110) argued for three conclusions:

(i) that the method of analysis is the fundamental element in Russell's philosophy;

(ii) that four distinctive applications of the method of analysis occur in Russell's philosophy: in ontology, in abstract cosmology, in mathematical logic and in the examination of the symbolism of science and ordinary life; and

(iii) that "Russell means by analysis a form of definition, either real definition of a non-Aristotelian sort, or contextual definition, i.e., definition of symbols in use."[3]

Russell in his "Reply" strongly endorsed Weitz's views, thereby ensuring their importance for later work (RTC: 684–6). Nevertheless, A.J. Ayer, rating Russell's "Reply" as "rather perfunctory", has argued that the Weitz view of Russellian analysis as a method of arriving at real or contextual definitions is "mistaken". Rather, asserts Ayer (1973: 10–11), the "central part" of Russellian analysis is the "ontological question" of "what there is or, rather, of what there can rationally be thought to be".[4] But has Ayer been entirely fair to Weitz? Weitz does not portray Russell as practising linguistic analysis for its own sake, as Ayer seems to think. Rather, Weitz's account of Russellian analysis has questions of ontology going hand in hand with questions of language. As Weitz repeatedly emphasises, Russellian analysis involves both contextual and real definition, where "contextual definition is concerned with symbols, real definition with *what* is symbolised" (1944: 120). So Ayer's difference of opinion with Weitz seems to come down to whether the "central part" of Russellian analysis deals with both symbols and what they symbolise or merely with what is symbolised. Put like this, the difference between the two positions is not as great as Ayer imagined. The issue will be clarified and resolved by a closer examination of what Russell had to say about analysis. It will turn out that he said a lot more on the subject than either Weitz or Ayer realised.

In the case of relations, there is a marked contrast between Russell's writings and those of his commentators. For instance, "relations(s)" occurs very frequently throughout *My Philosophical Development*, indicating the importance of this concept in *all* areas of philosophical interest to Russell. On the other hand the word is rarely found in writings *about* Russell's philosophy. This suggests that relations are seen as a technical device used by Russell in mathematical logic with limited relevance for his wider philosophy. This book will demonstrate that relations are used as a fundamental device throughout Russell's whole philosophy. It will be argued that an appreciation of the importance of both *analysis* and *relations* requires an elucidation of the close links between them. In brief, the claim will be that analysis typically *proliferates* relations.

If this link between analysis and relations is correct, then we can fruitfully turn our attention to Russell's theories of *space* and *time*. It will be argued that spatial and temporal relations serve as exemplars of relations in general. For various reasons Russell had to change his theories of space and time, this in turn altered his ideas on relations, thereby altering his philosophy as a whole. Not only were doctrines of space and time major concerns throughout his philosophical career, but they impinge on most of the major problems he investigated. From the early philosophy of geometry through epistemology, philosophy of mind and philosophy of science, Russell's theories of space and time and the changes they underwent, were crucial in shaping his distinctive philosophical conclusions. In the chapters that follow, it will emerge that these developments were indeed the result of the interplay of *analysis* and *relations*.

Finally a note on the method used to date the major phases of Russell's mature philosophy. It will be argued that, following his neo-Hegelian period, there are three such phases and they are dated from the *publication* of the first significant work which sets out the key ideas that characterise each phase. I am not denying that in some cases Russell may have entertained the ideas in the years before this first significant publication.

Notes

1. See, e.g., RTC: 700 and 705 and *MPD*: 11–12. That the misunderstanding of the later works has been rife is indicated by Denonn's Introduction to the 1954 second edition of *The Analysis of Matter*. Though convinced of the crucial importance of *The Analysis of Matter* and fulsome in its praise, Denonn makes two false claims: that Russell later abandoned unperceived events and that in *The Analysis of Matter* events are ultimate and unanalysable (cf. *AMa*: 286).
2. Grover Maxwell was a pioneer of this movement. See his essays in Pears (ed.), 1972 and in Nakhnikian (ed.), 1974. Savage and Anderson (eds.), 1989 is devoted to the continuity of Russell's early and late philosophy.
3. For a more recent account of Russell's method in mathematical philosophy that views it as "constructive definition" see Rodriguez-Consuegra, 1991, Section 5.3.
4. Honderich has given strong endorsement to Ayer's "persuasive and corrective view" (Honderich (ed.), 1984: 467).

PART ONE

Analysis and Relations – The Key to Continuity in Russell's Philosophy

CHAPTER 2

Russellian Analysis in Mathematical Philosophy

The part played by analysis in Russell's mathematical philosophy was stated clearly in a number of places in his writings. The task of mathematical philosophy is twofold. First,

> . . . to analyse existing mathematics, with a view to discovering what premisses are employed, whether these premisses are mutually consistent, and whether they are capable of reduction to more fundamental premisses.
>
> (*PM*, 'Preface': v).

In this way, analysis passes "from the complex to the simple, from the demonstrable to its indemonstrable premisses" (*PoM*: 3). This analysis "to greater and greater abstractness and logical simplicity" (*IMP*: 1) was viewed by Russell as the most philosophical part of mathematical philosophy. The second, more mathematical, part of the task of mathematical philosophy takes us in the opposite direction from analysis. When

> . . . we have decided upon our premisses, we have to build up again as much as may seem necessary of the data previously analysed, and as many other consequences of our premisses as are of sufficient general interest to deserve statement.
>
> (*PM*, 'Preface': v).

This synthesis stage is capable of yielding more than the existing mathematics that was analysed in the first place. According to Russell, we

> . . . shall find that by analysing our ordinary mathematical notions we acquire fresh insight, new powers, and the means of reaching whole new mathematical subjects by adopting fresh lines of advance after our backward journey.
>
> (*IMP*: 2).

Russell stresses a number of very important points about such analysis in mathematical philosophy:

(i) It is unlikely to be final.

We . . . "have no reason to suppose that it is impossible to find simpler ideas and axioms by means of which those with which we start could be defined and demonstrated" (*PM*, 'Preface': vi).[1]

(ii) As a corollary of (i), *the 'existing mathematics' on which analysis is practised changes as the subject itself evolves.*
As the frontiers are pushed back, yesterday's premisses become tomorrow's results from which a new generation of mathematical philosophers will start the backwards journey of analysis. Thus the mathematical philosophy/mathematics distinction "is one, not in the subject matter, but in the state of mind of the investigator" (*IMP*: 1).

(iii) The premisses established by analysis are far from being self-evident.
"In mathematics, the greatest degree of self-evidence is usually not to be found quite at the beginning, but at some later point; hence the early deductions, until they reach this point, give reasons rather for believing the premisses because true consequences follow from them, than for believing the consequences because they follow from the premisses" (*PM*, 'Preface': v).[2] A corollary of this, of course, is that there may be alternative premisses from which the same given set of results is deducible. This is the basis of Russell's characteristic open-mindedness about the finality or otherwise of his philosophical views at any given stage.

Russell's use of the terms 'premisses' and 'results' in his discussions of analysis requires comment. In the strict sense, of course, premisses and results, being components of deductive arguments, can only be *propositions* or *statements*. However, analysis leads not only to propositions, but also to *concepts* or *ideas* which are *primitive* at one level of analysis and *defined* at the next level down.[3] At the higher level these concepts or ideas are used in *definitions* that provide further premisses. When characterising his method of analysis, Russell sometimes, for convenience, uses 'premisses' in a *wider sense* to refer to concepts or ideas, as well as strict propositions. Take, for instance, Peano's analysis of natural number theory via three primitive concepts and five primitive propositions. In Russell's wider sense the three concepts and five propositions are the premisses, yet strictly speaking the only premisses are the five primitive propositions. However, including the concepts (0, number and successor) amongst the premisses is fairly innocuous since they are used to *define* other premisses (e.g. "1 is the successor of 0"). In the next breakthrough in analysis, due to Frege, the concepts ceased to be primitive (i.e., he provided a definition of number). This wider sense of 'premisses' is typically employed in Russell's descriptions of philosophical analysis.

On the point about the premisses being not self-evident, Russell was sharply critical of Leibniz. Whilst approving of Leibniz's conception of a 'Universal Mathematics' embracing formal logic as well as algebra, geometry, etc., he rejected Leibniz's view that the basic premisses of such a general calculus would be 'identical', i.e. analytic. This error of Leibniz leads to a mistaken view of philosophy:

> . . . the business of philosophy is just the discovery of those simple notions, and those primitive axioms, upon which any calculus or science must be based. The belief that the primitive axioms are identical leads to an

> emphasis on *results*, rather than premisses, which is radically opposed to the true philosophic method. There can be neither difficulty nor interest in the premisses, if these are of such a kind as 'A is A' or 'AB is not non-A'. And thus Leibniz supposed that the great requisite was a convenient method of deduction. Whereas, in fact, the problems of philosophy should be anterior to deduction. An idea which can be defined, or a proposition which can be proved, is of only subordinate philosophical interest. The emphasis should be laid on the indefinable and indemonstrable, and here no method is available save intuition.
>
> (*PL*: 170–1).

The 'simple notions' and 'primitive axioms' which Russell here claims it is the business of philosophy to discover, are equivalent to the *concepts* and *propositions* of the discussion above. Even if the *propositions* are analytic, there is still scope for further analysis of the *concepts* that are taken as primitive at that level of analysis. This is why Russell's later acceptance of Wittgenstein's view "that logic consists wholly of tautologies" (*MPD*: 88)[4] in no way represents the end of the "backward journey into the logical foundations of mathematics" (*IMP*: 204–5).[5] The response to Wittgenstein is essentially the same as the one to Leibniz almost twenty years earlier.

The accuracy of the preceding account of analysis in Russell's mathematical philosophy[6] is readily confirmed by his consideration of the case of the natural numbers.[7] (In the following summary, 'premisses' is used in the wider sense noted above, i.e., propositions and concepts.) For the Greeks and Romans, the natural numbers were 1, 2, 3, 4, . . . etc. At a later stage the frontiers were pushed back so that they became 0, 1, 2, 3, 4, . . . n, n +l, . . . , etc., with 0 being a far from self-evident addition to the premisses. Increasingly it started to look as though the natural numbers themselves were the premisses from which all of traditional pure mathematics could be deduced, with incommensurability being the chief obstacle to this view.

Peano contributed the next step in analysis by discovering a new set of premisses (three primitive ideas and five primitive propositions) from which, with the help of logic, the theory of natural numbers could be deduced. However one aspect of Peano's premisses not being self-evident is that they admit of a plurality of interpretations. The next stage in analysis was the 'logicisation of arithmetic', i.e., from 'premisses' belonging solely to logic, Peano's three primitive ideas could be defined and his five primitive propositions deduced. This work is largely due to Frege, Whitehead and Russell.

Throughout this development, the features of analysis considered above are evident:

(i) Analysis is unlikely to be final.

The logicist program of *Principia Mathematica* was never perfectly realised. The point, however, is that even if it had been it would not have ruled out the possibility of yet further analysis.

(ii) *Analysis enlarges our understanding of mathematics as the frontier is pushed back.*
The logic of relations, e.g., becomes part of mathematics.

(iii) *Analysis results in premisses becoming less and less self-evident.*
E.g. a "number is anything which is the number of some class", where the "number of a class is the class of all those classes that are similar to it" (*IMP*: 18–9) is clearly a less self-evident definition than "a number is any of 1, 2, 3, 4, . . . etc." (These definitions count as premisses in the wider sense above.)

Before proceeding to the role of analysis in philosophy in general, two points about the origin of Russell's technique of analysis will be considered. First, Russell's view that analysis is the appropriate method for philosophy appears to be inspired by the role that analysis plays in mathematics. Analysis as a mathematical method involves commencing with what is to be proved (i.e., the conclusion) and moving backwards to some fundamental truths or axioms.[8] (By contrast, synthesis combines basic truths or premisses to deduce a conclusion.) Generally speaking, Greek mathematics (including, of course, Euclidean geometry) is *synthetic* in its methods. Thus Euclid's *Elements* lists in the first book three kinds of assumptions: definitions, postulates and common notions. The procedure then is to prove and solve propositions and problems using assumptions of this sort. We are inevitably reminded of Aristotle's ideal of a deductive science. On the other hand, modern mathematics, stemming from 17th century developments in algebra and calculus is generally analytic.[9] Whilst the Greek synthetic method had the effect of enshrining common notions in mathematics, the modern analytic approach is more likely to dissolve common notions, as we have seen already.

Second, examples of analysis in mathematical philosophy that conform with the account developed above are to be found in Russell's idealist phase. Thus, whereas Russell and Moore started the movement that became known as 'analytic philosophy' somewhere around 1899, the *analytic method* predates that. While the continuities between Russell's method in his idealist phase and in his later phases will be discussed more fully in Section 4.2.3, the early influence of Kant will be considered here. Russell's fellowship dissertation (*EFG*) is an attempt to refurbish Kant's views in the light of non-Euclidean geometry, and, amongst Kant's methods of argument there is a definite prefiguring of Russellian analysis.

Russell found in Kant "two classes of arguments" concerning space and geometry (*EFG*: lff., 55ff).[10] The first class are "transcendental" (or "epistemological" or "philosophical") arguments and conclude (e.g.) that space must be subjective and *a priori* as the only possible explanation of the premiss that Euclidean geometry is known to have apodeictic certainty. Another example of this class of argument is, of course, Kant's famous transcendental deduction in the 'Analytic of Concepts' section of the *Critique of Pure Reason*. The transcendental argument to the nature of space is refuted by the possibility of non-Euclidean geometries, since these show that it is not the case that

Euclidean geometry is known to have apodeictic certainty. More exactly, we now distinguish between geometries as formal systems and geometry as a branch of physics, the former being *a priori* but not synthetic, the latter synthetic but not *a priori*. There is no synthetic *a priori* geometry to serve as the starting point for a transcendental argument. Russell's reaction to this in *EFG* was to try to isolate those axioms which are *a priori* necessary to any geometry and deduce from these the *a priori* necessary properties of any "form of externality". This attempt by Russell to refurbish Kant's transcendental argument by starting with a weaker certainty and then following a similar procedure to his illustrious predecessor *might* be described as analysis in some senses of that wide term. However, as we will find, this is *not* the sort of analysis that was to become distinctively Russellian.

The second class of arguments that Russell finds in Kant are "metaphysical" (or "geometrical") arguments and these are the ones that point the way to the Russellian method of analysis. In Kant, these arguments involve a general philosophical analysis of the concepts under consideration. Thus, in the Transcendental Aesthetic Kant analyses the most general features of space and time and attempts to deduce their *a priori* and subjective character. Russell wasn't very impressed with these arguments of the Transcendental Aesthetic (*EFG*: 1),[11] and instead developed his own "geometrical argument" as a replacement for Kant's efforts. This "geometrical argument" was seen by Russell as providing a solid basis for geometry and represents a clear early example of his method of analysis. Before examining Russell's "geometrical argument", it is noteworthy that Kant's metaphysical deduction in the Analytic of Concepts is a good example of one of his metaphysical arguments anticipating Russell's method of analysis. What Kant attempts there is to set out the major formal features of thinking anything at all (Russell's 'premisses') and from these to deduce the major formal features of thinking about objects of experience (Russell's 'results'). That not many people have thought him successful is beside the point.

How then is Russell's idealist "geometrical argument" an application of the analytic method? What the argument undertakes is an "analysis of geometrical reasoning" (APG: 292ff., *EFG*: Ch. III). This entails starting with a result (or conclusion) and working backwards to fundamental premisses from which the result can be deduced. The 'result' in this case is the thesis that geometry is the science of *spatial magnitude*, i.e., of the measurement of space. The three axioms (of free mobility, of dimensions and of the straight line) are found by analysis to be the premisses that jointly imply that measurement of space is possible.

At a time when few idealists would admit that any geometry other than Euclidean was possible, Russell had laboured for years to develop an account of geometry that was compatible with idealist thinking, whilst being based on contemporary developments in mathematics. It wasn't to be long before he became convinced that idealist presuppositions created intractable difficulties for the elucidation of mathematics. By discarding his allegiance to idealism, he was able to adapt his analytic method to some very recent developments in mathematical philosophy and to make real progress. In the process he ran into a variety of new difficulties that were to preoccupy him for more than a decade.

NOTES

1. Cf. *IMP*: 3–4.
2. See also LA: 164, PIML: 33 and *AMa*: 2.
3. See, e.g., *IMP*: 3–4.
4. For fuller discussion, see MPL.
5. *AMa*: 172 gives a different reason for supporting this conclusion.
6. See Irvine (1989) for a discussion of Russell's method in mathematical philosophy and its relation to logicism that agrees substantially with my account. A more recent account of Russell's turn of the century method in mathematical philosophy (Rodriguez-Consuegra, 1991, Section 5.3) views it as "constructive definition". According to Rodriguez-Consuegra, analysis defines concepts in simpler terms which then enables the concept to be constructed "out of materials belonging to lower ontological levels" (1991: 205).
7. The brief account that follows is based on *IMP*: Chs. I–III. See also RMDP: 275–82 and *AMa*: 3–4.
8. See, e.g., J. Bendick and M. Levin (1971) *Mathematics Illustrated Dictionary* (London: Kaye & Ward) and G. James and R.C. James (1976) *Mathematical Dictionary* (4th edn.) (New York: Van Nostrand Reinhold Co.).
9. For an interesting discussion of these trends see Torretti (1978, Ch. 1).
10. Also APG and *HWP*: 685ff. Kant also applies the same sorts of arguments to time and arithmetic.
11. Also *HWP*: 685ff.

CHAPTER 3

Russellian Analysis in General Philosophy

3.1 Russell's Characterisations of Analysis

As pointed out in Chapter One, there is a common view that Russell practised analysis whilst failing to detail what it involved. Copleston, e.g., agrees with Weitz that there are "methods of analysis" in Russell, rather than a single method (1967: 204 and fn. 46). Whilst it is true that the topic never received from Russell the extended treatment that posterity would desire, nevertheless he did leave some important pointers. For a start it is clear that analysis in mathematical philosophy is the model for all of Russell's philosophical analysis.[1]

> The business of philosophy, as I conceive it, is essentially that of logical analysis, followed by logical synthesis.
>
> (LA: 176).

Logical analysis in general philosophy involves the same backwards journey as it did in mathematical philosophy:

> . . . every truly philosophical problem is a problem of analysis; and in problems of analysis the best method is that which sets out from results and arrives the premisses.
>
> (PIML: 33).

Likewise general philosophy follows mathematical philosophy in having logical analysis mirrored by logical synthesis:

> When the philosopher's work has been perfectly accomplished, its results can be wholly embodied in premisses from which deduction may proceed.
>
> (*PoM*: 129).

Russell also thought that his method of analysis was something new in philosophy:

> The impulse to philosophy remains very strong in me — I feel that I really have got a method that gives more precision than there has ever been before, and more power of getting at the skeleton of the world — the framework that things are built on. It is difficult to get people to see it, because philosophers

are not trained to precision. . . .

(Letter of 29 October 1912, quoted in *Papers* 6, Introduction)

Hence I conclude that we have a *prima facie* case for supporting the thesis that there is a single method of analysis throughout all Russell's philosophy, whether mathematical or otherwise.

Each of the last four quotations comes from a context where Russell is emphatically asserting that his method of analysis has general application in philosophy. Likewise, when summing up his career, Russell repeatedly stated that a *single* method was common to all of his philosophical ventures (*MPD*: 98, 162).[2] Hence we would expect that the three features of analysis, noted in the case of mathematical philosophy, would turn out to apply to Russellian analysis in general. In what follows it will be shown that this is so, thereby challenging some prevailing interpretations of Russell's work.

Someone might object to the 'single method of analysis' claim as follows. Since the above quotations are from works of widely different periods in Russell's career (1903 for *PoM*, 1913 for PIML and 1924 for LA), might they not be selective in that other statements that he makes about analysis possibly don't support the claim. This objection will be met in two ways. The next chapter of this book will demonstrate how Russell's philosophy at all stages does indeed exemplify the account of analysis outlined here. However, in the short term the objection will be countered by a consideration of a wide sample of Russell's comments on analysis with the aim of showing that they all support the present account.

Here are the Russell quotations in chronological order:

A. (1904)

The method of philosophy is not fundamentally unlike that of other sciences: the differences seem to be only in degree. The data are fewer, but are harder to apprehend; and the inferences required are probably more difficult than in other subjects except mathematics. But the important point is that, in philosophy as elsewhere, there are self-evident truths from which we must start, and that these are discoverable by the process of inspection or observation, although the material to be observed is not, for the most part, composed of existent things.

(MTCA: 22–3).

B. (1910)

The process to be gone through is essentially one of analysis: we have various complex and more or less confused beliefs about the true and the false, and we have to reduce these to forms which are simple and clear, without causing any avoidable conflict between our initial complex and confused beliefs and our final simple and clear assertions. These final assertions are to be tested partly by their intrinsic evidence, partly by their power of accounting for the 'data'; and the 'data', in such a problem, are the complex and confused beliefs with which we start. These beliefs must necessarily suffer a change in becoming clear, but the change should not be greater than is warranted by their initial confuson.

(NTF: 116–7).

C. (1914)

Before embarking upon our analysis, let us again take stock of those relevant facts which are least open to doubt. From the diversity of philosophical theories on the subject, it is evident that the true analysis, whatever it may be, cannot itself be among the facts that are evident at once, but must be reached, like a scientific hypothesis, as the theoretic residue left by the comparison of data. Here as in philosophy generally, it is not the few logically simplest facts that form our data, but a large mass of complex everyday facts, of which the analysis offers fresh difficulties and doubts at every step. For this reason, if we wish to start with what is undeniable, we have to use words, at first, which, though familiar, stand in need of a dissection and definition only possible at a later stage.

(ONA: 160).[3]

D. (1914)

In every philosophical problem, our investigation starts from what may be called 'data', by which I mean matters of common knowledge, vague, complex, inexact, as common knowledge always is, but yet somehow commanding our assent as on the whole and in some interpretation pretty certainly true. . . . We are quite willing to admit that there may be errors of detail in this knowledge, but we believe them to be discoverable and corrigible by the methods which have given rise to our beliefs, and we do not, as practical men, entertain for a moment the hypothesis that the whole edifice may be built on insecure foundations. In the main, therefore, and without absolute dogmatism as to this or that special portion, we may accept this mass of common knowledge as affording data for our philosophical analysis.

It may be said – and this is an objection which must be met at the outset – that it is the duty of the philosopher to call in question the admittedly fallible beliefs of daily life, and to replace them by something more solid and irrefragable. In a sense this is true, and in a sense it is effected in the course of analysis. But in another sense, and a very important one, it is quite impossible. While admitting that doubt is possible with regard to all our common knowledge, we must nevertheless accept that knowledge in the main if philosophy is to be possible at all. There is not any superfine brand of knowledge, obtainable by the philosopher, which can give us a standpoint from which to criticise the whole of the knowledge of daily life.

The most that can by done is to examine and purify our common knowledge by an internal scrutiny, assuming the canons by which it has been obtained, and applying them with more care and with more precision. Philosophy cannot boast of having achieved such a degree of certainty that it can have authority to condemn the facts of experience and the laws of science. The philosophic scrutiny, therefore, though sceptical in regard to every detail, is not sceptical as regards the whole. That is to say, its criticism of details will only be based upon their relation to other details, not upon some external criterion which can be applied to all the details equally. The reason for this abstention from a universal criticism is not any dogmatic confidence, but its exact opposite; it is not that common knowledge *must* be true, but that we possess no radically different kind of knowledge derived from some other source. Universal scepticism, though logically irrefutable, is practically barren; it can only, therefore, give a certain flavour of hesitancy to our beliefs, and cannot be used to substitute other beliefs for them.

(*OKEW2*: 72–4).[4]

E. (1914)

Every subject-matter, it would seem, can give rise to philosophical investigations as well as to the appropriate science, the difference between the two treatments being in the direction of movement and in the kind of truths which it is sought to establish. In the special sciences, when they have become fully developed, the movement is forward and synthetic, from the simpler to the more complex. But in philosophy we follow the inverse direction: from the complex and relatively concrete we proceed towards the simple and abstract by means of analysis, seeking, in the process, to eliminate the particularity of the original subject-matter, and to confine our attention entirely to the logical *form* of the facts concerned . . . philosophy, starting from data which are common knowledge, seeks to purify and generalise them into the simplest statements of abstract form that can be obtained from them by logical analysis.

(*OKEW2*: 189–90).

F. (1914)

The nature of philosophic analysis . . . can . . . be stated in general terms. We start from a body of common knowledge, which constitutes our data. On examination, the data are found to be complex, rather vague, and largely interdependent logically. By analysis we reduce them to propositions which are as nearly as possible simple and precise, and we arrange them in deductive chains, in which a certain number of initial propositions form a logical guarantee for all the rest. These initial propositions are *premisses* for the body of knowledge in question. Premisses are thus quite different from data – they are simpler, more precise, and less infected with logical redundancy. If the work of analysis has been performed completely, they will be wholly free from logical redundancy, wholly precise, and as simple as is logically compatible with their leading to the given body of knowledge. The discovery of these premisses belongs to philosophy; but the work of deducing the body of common knowledge from them belongs to mathematics, if 'mathematics' is interpreted in a somewhat liberal sense.

But besides the logical analysis of the common knowledge which forms our data, there is the consideration of its degree of certainty. When we have arrived at its premisses, we may find that some of them seem open to doubt, and we may find further that this doubt extends to those of our original data which depend upon these doubtful premisses. . . . We may say, generally, that what commonly passes as knowledge is not all equally certain, and that, when analysis into premisses has been effected, the degree of certainty of any consequence of the premisses will depend upon that of the most doubtful premiss employed in proving this consequence. Thus analysis into premisses serves not only a logical purpose, but also the purpose of facilitating an estimate as to the degree of certainty to be attached to this or that derivative belief. In view of the fallibility of all human beliefs, this service seems at least as important as the purely logical services rendered by philosophical analysis.

(*OKEW2*: 214–5).

G. (1918)

It is a rather curious fact in philosophy that the data which are undeniable to start with are always rather vague and ambiguous. You can, for instance, say: 'There are a number of people in this room at this moment.' That is obviously in some sense undeniable. But when you come to try and define what this room is, and what it is for a person to be in a room, and how you are going to distinguish one person from another, and so forth, you find that what you have said is most fearfully vague and that you really do not know what you mean. That is a rather singular fact, that everything you are really sure of, right off is something that you do not know the meaning of, and the moment you get a precise statement you will not be sure whether it is true or false, at least right off. The process of sound philosophising, to my mind, consists mainly in passing from those obvious, vague, ambiguous things, that we feel quite sure of, to something precise, clear, definite, which by reflection and analysis we find is involved in the vague thing that we start from, and is, so to speak, the real truth of which that vague thing is a sort of shadow.

(PLA: 161).

H. (1924)

... I began to think it probable that philosophy had erred in adopting heroic remedies for intellectual difficulties, and that solutions were to be found merely by greater care and accuracy. This view I have come to hold more and more strongly as time went on, and it has led me to doubt whether philosophy, as a study distinct from science and possessed of a method of its own, is anything more than an unfortunate legacy from theology.

(LA: 163).

I. (1924)

The technical methods of mathematical logic, as developed in [*Principia Mathematica*] seem to me very powerful, and capable of providing a new instrument for the discussion of many problems that have hitherto remained subject to philosophic vagueness. ...

When pure mathematics is organised as a deductive system – i.e. as the set of all those propositions that can be deduced from an assigned set of premisses – it becomes obvious that, if we are to believe in the truth of pure mathematics, it cannot be solely because we believe in the truth of the set of premisses. Some of the premisses are much less obvious than some of their consequences, and are believed chiefly because of their consequences. This will be found to be always the case when a science is arranged as a deductive system. It is not the logically simplest propositions of the system that are the most obvious, or that provide the chief part of our reasons for believing in the system. With the empirical sciences this is evident. Electro-dynamics, for example, can be concentrated into Maxwell's equations, but these equations are believed because of the observed truth of certain of theirlogical consequences. Exactly the same thing happens in the pure realm of logic; the logically first principles of logic – at least some of them – are to be believed, not on their own account, but on account of their consequences.

(LA: 163–4).

J. (1924)

The business of philosophy, as I conceive it, is essentially that of logical analysis, followed by logical synthesis. Philosophy is more concerned than any special science with relations of different sciences and possible conflicts between them; in particular, it cannot acquiesce in a conflict between physics and psychology, or between psychology and logic. Philosophy should be comprehensive, and should be bold in suggesting hypotheses as to the universe which science is not yet in a position to confirm or confute. But these should always be presented *as* hypotheses, not (as is too often done) as immutable certainties like the dogmas of religion. Although, moreover, comprehensive construction is part of the business of philosophy, I do not believe it is the most important part. The most important part, to my mind, consists in criticising and clarifying notions which are apt to be regarded as fundamental and accepted uncritically. As instances I might mention: mind, matter, consciousness, knowledge, experience, causality, will, time. I believe all these notions to be inexact and approximate, essentially infected with vagueness, incapable of forming part of any exact science. Out of the original manifold of events, logical structures can be built which will have properties sufficiently like those of the above common notions to account for their prevalence, but sufficiently unlike to allow a great deal of error to creep in through their acceptance as fundamental.

(LA: 176–7).

K. (1925)

It is a curious fact – of which relativity is not the only illustration – that, as reasoning improves, its claims to the power of proving facts grow less and less. Logic used to be thought to teach us how to draw inferences; now, it teaches us rather how not to draw inferences

(*ABCR*: 155–6).

L. (1944)

As a general rule, the effect of logical analysis is to show the mutual independence of propositions which had been thought to be logically connected. . . . As logic improves, less and less can be proved . . . The result of logical analysis is to increase the number of independent premisses that we accept in our analysis of knowledge.

(RTC: 684).

M. (1944)

Modern analytical empiricism, of which I have been giving an outline, differs from that of Locke, Berkeley, and Hume by its incorporation of mathematics and its development of a powerful logical technique. It is thus able, in regard to certain problems, to achieve definite answers, which have the quality of science rather than of philosophy. It has the advantage, as compared with the philosophies of the system-builders, of being able to tackle its problems one at a time, instead of having to invent at one stroke a block theory of the whole universe. Its methods, in this respect, resemble those of science. I have no doubt that, in so far as philosophical knowledge is possible, it is by such methods that it must be sought; I have also no doubt that, by these methods, many ancient problems are completely soluble.

There remains, however, a vast field, traditionally included in philosophy, where scientific methods are inadequate. This field includes ultimate questions of value; science alone, for example, cannot prove that it is bad to enjoy the infliction of cruelty. Whatever can be known, can be known by means of science; but things which are legitimately matters of feelings lie outside its province . . . philosophers who make logical analysis the main business of philosophy . . . confess frankly that the human intellect is unable to find conclusive answers to many questions of profound importance to mankind, but they refuse to believe that there is some 'higher' way of knowing, by which we can discover truths hidden from science and the intellect. For this renunciation they have been rewarded by the discovery that many questions, formerly obscured by the fog of metaphysics, can be answered with precision, and by objective methods which introduce nothing of the philosopher's temperament except the desire to understand. Take such questions as: What is number? What are space and time? What is mind, and what is matter? I do not say that we can here and now give definitive answers to all these ancient questions, but I do say that a method has been discovered by which, as in science, we can make successive approximations to the truth, in which each new stage results from an improvement, not a rejection, of what has gone before.

In the welter of conflicting fanaticisms, one of the few unifying forces is scientific truthfulness, by which I mean the habit of basing our beliefs upon observations and inferences as impersonal, and as much divested of local and temperamental bias, as is possible for human beings. To have insisted upon the introduction of this virtue into philosophy, and to have invented a powerful method by which it can be rendered fruitful, are the chief merits of the philosophical school of which I am a member.

(*HWP*: 788–9).

N. (1959)

My method invariably is to start from something vague but puzzling, something which seems indubitable but which I cannot express with any precision. I go through a process which is like that of first seeing something with the naked eye and then examining it through a microscope. I find that by fixity of attention divisions and distinctions appear where none at first was visible, just as through a microscope you can see the bacilli in impure water which without the microscope are not discernible. There are many who decry analysis, but it has seemed to me evident, as in the case of the impure water, that analysis gives new knowledge without destroying any of the previously existing knowledge. This applies not only to the structure of physical things, but quite as much to concepts. 'Knowledge', for example, as commonly used is a very imprecise term covering a number of different things and a number of stages from certainty to slight probability.

It seems to me that philosophical investigation, as far as I have experience of it, starts from that curious and unsatisfactory state of mind in which one feels complete certainty without being able to say what one is certain of. The process that results from prolonged attention is just like that of watching an object approaching through a thick fog: at first it is only a vague darkness, but as it approaches articulations appear and one discovers that it is a man or a woman, or a horse or a cow or what not. It seems to me that those who object to analysis would wish us to be content with the initial dark blur. Belief in the above process is my strongest and most unshakable prejudice as regards the methods of philosophical investigation.

(*MPD*: 98–9).

O. (1959)

What I myself have had to say, whether about mathematics or about physics or about perception or about the relation of language to fact, has proceeded always by a certain method. Taking it for granted that, broadly speaking, science and common sense are capable of being interpreted so as to be true in the main, the question arises: what are the minimum hypotheses from which this broad measure of truth will result? This is a technical question and it has no unique answer. A body of propositions, such as those of pure mathematics or theoretical physics, can be deduced from a certain apparatus of initial assumptions concerning initial undefined terms. Any reduction in the number of undefined terms and unproved premisses is an improvement since it diminishes the range of possible error and provides a smaller assemblage of hostages for the truth of the whole system.

(*MPD*: 162).

P. (1959)

The relation of empirical knowledge to experience is one of those problems best treated by beginning with something vague but indubitable and proceeding by scrutiny to prove that the vague something involves something much more precise but *prima facie* much less indubitable.

(*MPD*: 168).

Q. (1959)

All the advances of modern physics have consisted in a more and more minute analysis of the material world. Atoms at first were considered almost unbelievably small, but to the modern physicist each atom is a complicated world like the solar system. . . . It is not only in relation to matter that analysis is the road to understanding. A person without musical training, if he hears a symphony, acquires a vague general impression of a whole, whereas the conductor, as you may see from his gestures, is hearing a total which he minutely analyses into its several parts. The merit of analysis is that it gives knowledge not otherwise obtainable. When you learn that water consists of two parts of hydrogen to one of oxygen, you do not cease to know anything that you knew before about water, but you do acquire the power to know many things that an un-analytic observation could not teach you.

(*MPD*: 169–70).

The foregoing, then, is a wide representative sample of Russell's statements about philosophical analysis. In the discussion that now follows, the various quotations will be referred to by their respective capital letters. Before drawing out the common threads from this series of quotations and showing how they support our account of Russellian analysis, a couple of cautionary points:

a) At first glance some of the quotations appear to contradict others, e.g. the claim in A that "there are self-evident truths from which we must start" appears to clash with the notion in I that "some of the premisses are much less obvious than some of their consequences". The appearance of contradiction is dispelled by noting two points. Firstly, there are, as we have already found, two halves to Russellian analysis, viz. *logical analysis* and *logical synthesis*. Analysis starts at the results and moves to the premisses, a movement Russell often refers to as the *inductive* order or "order of knowledge",[5] since this is the direction historically of knowledge advance. Synthesis, of course, starts at the premisses and moves by deduction to the results thereby embodying the *deductive* or logical order. Thus the "self-evident truths" from which A claims "we must start" are the *results*, hence A is consistent with I. The second point to note is that though the results (as compared with the premisses) are "self-evident", "undeniable" and "indubitable", they are also "inexact", "vague" and "confused". These two sets of characteristics are compatible as Russell's examples in G, N and Q show.
b) While most of the quotations are clearly concerned with the features of Russellian analysis discussed previously, the relevance of some, e.g. H and K, is not obvious yet. Their importance will become apparent in subsequent discussion.

3.2 Further Elaboration of Russellian Analysis

To what extent, then, do these quotations support our previous account of Russellian analysis? For a start, M and especially N and O, coming in passages that sum up his career, clearly show Russell's belief that a *single* method was common to *all* of his ventures, mathematical or otherwise. It is equally clear that that method was the analysis of results to arrive at premisses. Likewise, the three features of analysis highlighted in the previous chapter on mathematical philosophy figure prominently in these quotations:

(i) *Analysis unlikely to be final.*
The lack of certainty of many philosophical premisses (F), philosophy's role in suggesting possible hypotheses (J), and the fact that it is a "technical question" with "no unique answer" as to what is the minimum set of premisses (O) – all of these are obviously related to the non-finality of analysis.

(ii) *Analysis enlarges our understanding of particular subjects.*
Just as we saw philosophical analysis extending our understanding of the nature of mathematics, so also with science (cf. E, J, Q). Indeed, this role may exhaust the usefulness of philosophy (H), i.e. Russell suggests that all sound philosophy may be parasitic on science and mathematics. This view locates philosophy at the frontiers of the particular exact sciences. As the frontier is extended, territory that once belonged to philosophy becomes exact enough to be incorporated into science. Thus "every advance in knowledge robs philosophy of some problems which formerly it had. . . . " (PLA: 243). It remains for philosophy to move to the new frontier. Hence Russell's description of philosophy as occupying the "No Man's Land" between "theology and science" (*HWP*: 13) and the maxim that "science is what you more or less know and philosophy is what you do not know" (PLA: 243).

(iii) *Analysis leads to premises that are decreasingly self-evident.*
This point is emphatically made in I and P, however we need to beware of a certain confusion that results from ambiguity of the word 'simple' and the use Russell makes of it on different occasions. He variously describes the results and premisses as follows:

Table 3.1. Characteristics of Russellian results and premisses

Results (or data)	Premisses
More complex (B, C, D, E, F)	Simpler (B, C, E, F, I)
Relatively concrete (E)	Abstract (E)
Common knowledge (D, E, F)	
Vague (D, F, G, J, N, P, Q)	Precise (F, G, P)
Logically interdependent (F, L)	Logically independent (F, L)
More obvious (G, I)	Less obvious (I)
Undeniable (C, G)	
Inexact and approximate (D, J)	Definite (G)
Indubitable (N, P)	Dubitable (K, P)
Puzzling (N)	
Confused (B)	Clear (B, G)
Self-evident (A)	
Ambiguous (G)	

This greatly assists in our understanding of the claim that the premisses are less self-evident than the results. The premisses are *simple* in the primary sense that the results can be *compounded* from them. However, as the Oxford dictionary confirms, 'simple' also means 'easily understood', i.e., the *results* could also be seen as simple in that they are concrete, common knowledge, obvious and indubitable. Russell appears to have been using the term in this second sense when he said that "the point of philosophy is to start with something so simple as not to seem worth stating, and to end up with something so paradoxical that no one will believe it" (PLA: 172). This interpretation makes Russell's statement perfectly consistent with the above account of analysis.

However, there is an even more fundamental reason why there is confusion about simples in Russell's philosophy. It stems, I believe, from another ambiguity – this time in what Russell means by 'analysis'. The reader has likely noticed already that on one understanding of the term *analysis* is the first, and more philosophical, stage of Russell's method. The second, more mathematical or logical, stage is, of course, synthesis. However, on the other understanding, *analysis* is the name of Russell's philosophical method. Let us call the former understanding the *narrow* interpretation of analysis, and the latter the *broad* interpretation. I suggest that the confusion resulting from these two meanings of 'analysis' has led people to concentrate on the first stage of Russell's philosophical method and treat that as all there is to it. Hence, e.g., Sainsbury's claim that analysis is "reducing complex truths to simpler truths, and complex things to simpler things" (1979: 235). Of course, this is fairly accurate as a description of the first stage, but Sainsbury, along with most others, thinks it is the whole story. Which, as we have shown, it is not. What is left out makes all the difference about how one treats relations in Russell's philosophy, as we will find later.

Another important aspect of Russell's philosophy that needs to be reconciled with Table 3.1 is the principle of acquaintance. How does the analysis of relatively concrete data into relatively abstract premisses (as emphatically shown in Table 3.1) fit with Russell's commitment to the principle of acquaintance? The answer is that typical objects of acquaintance, such as sense-data and relations, are more abstract than the relatively concrete data (such as chairs and tables) that they displace. As will be shown in Chapter Seven, while a logical construction is itself a created abstract entity, its elements are all entities that are known by acquaintance. The logical construction serves the same function as the inferred entity that it displaces, e.g., in explaining data. It is more secure epistemically because all of the elements of the construction are objects of acquaintance. So, while the logical construction is a new abstract complex, its elements are types of particulars and relations previously known by acquaintance.

A determined sceptic might acknowledge that the preceding selection of quotations does indeed demonstrate a common thread in various Russellian characterisations of analysis, yet nevertheless hold that the common thread is overlaid by important differences at various crucial stages in Russell's work. These differences, so the sceptic would urge, are substantial enough to lead to analysis at such stages being distinctly unlike analysis at others. Hence the above unified account would be liable to overlook important changes in Russell's method.

Certainly it is not difficult to find accounts of Russell's work that have him departing radically from the method of analysis outlined above. To consider but two examples:

a) Lakatos views Russell's logicism in mathematical philosophy as an attempt to deduce all of mathematics from trivial, indubitable logical premisses, whereas our account holds that the premisses are less obvious and more dubitable than the results deduced from them.

b) Many writers on Russell's logical atomism claim that basic statements about sense-data are the premisses. Such statements are held by Russell to be self-evident, yet our account has analysis reaching premisses that are decreasingly self-evident.

The defence of our thesis against challenges such as these will proceed in three stages:

(i) A more detailed account of Russellian analysis will be offered, further explicating the differences between the narrow and broad interpretations encountered above.

(ii) It will be shown that Russell subscribed to this detailed account at each major phase in his career. In the process the contrary views expressed in (a) and (b) will be examined closely and rejected.

(iii) It will be shown that famous instances of Russell's philosophising, such as the theory of descriptions and the various logical constructions, conform to our detailed account.

(i) and (ii) will occupy the next chapter, while (iii) will wait till later chapters.

Notes

1. Russell often states that his work in non-mathematical philosophy is founded on applications of the method of analysis used in mathematical philosophy. See, e.g., *OKEW*, Preface and PLA, first few pages. Warnock (1969: 33–6) stresses this point.
2. See also *HWP*: 788–9.
3. See also *Papers* 7: 33.
4. This and the following two quotations were not altered by Russell in 1926 when he revised *OKEW* in the light of his changed philosophical views.
5. See, e.g., RMDP: 282 and PIML: 40.

CHAPTER 4

A Systematic Account of Russellian Analysis

4.1 Analysis Diagrams

The following diagram is a schematic representation of Russellian analysis. (Hereafter such diagrams will be called 'analysis diagrams'.) The diagram incorporates major features of Russellian analysis discussed above, such as the movement from common knowledge (*results*) to abstract propositions (*premisses*), followed by the return journey to propositions which capture the truths of the common knowledge whilst avoiding the errors of the latter (*analysed results*). The earlier distinction we made between narrow and wide interpretations of analysis is represented, respectively, as *logical analysis* and *philosophical analysis*.

Of course an analysis diagram doesn't incorporate *all* of the important features of Russellian analysis discussed earlier. For one thing, no account has been taken of the non-finality of the premisses, i.e. of the point that today's *premisses* will very likely serve as tomorrow's *results* for the next step in pushing back the frontiers of knowledge. However, this feature *could* be included by making the analysis diagram more complex than it needs be for present purposes. Likewise the diagram doesn't incorporate Russell's point about the different degrees of certainty of the various premisses being transmitted to the analysed results. Once again this feature *could* be taken into account by increasing the complexity of the diagram. However, the analysis diagram in its present form *is* sufficient to support two very important points about applications of Russellian analysis in general.

4.1.1 Features of Russellian Analysis

Firstly, Russellian analysis is *primarily* analysis of propositions and only *indirectly* is it analysis of objects, i.e. analysis (and synthesis) occur only in the left hand half of the analysis diagram. According to Russell, the notion that "all sound philosophy should begin with an analysis of propositions, is a truth too evident, perhaps, to demand a proof" (*PL*: 8).[1] Thus, in his famous analysis of the desk (PLA: 236ff.), it is not the desk that is analysed, but rather propositions

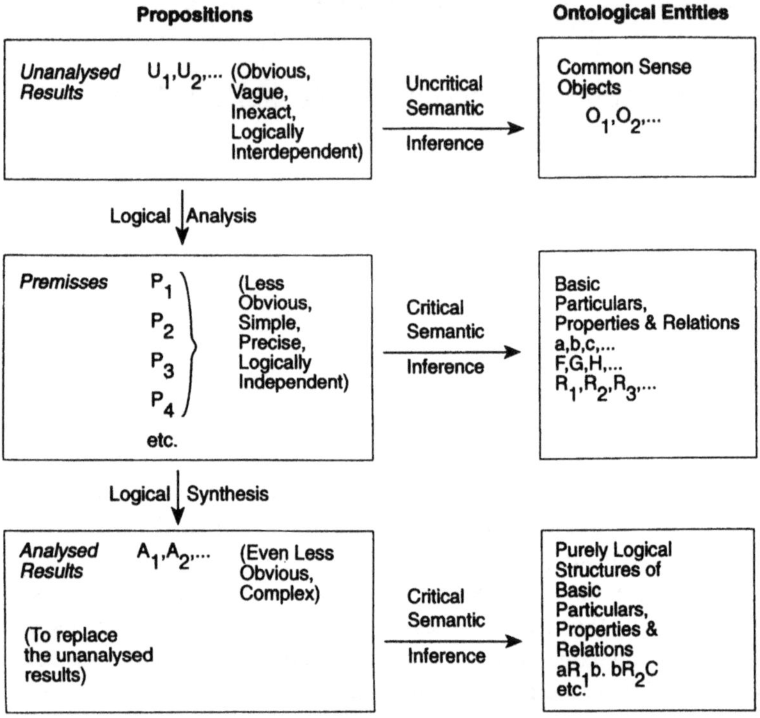

Figure 4.1. Russellian Philosophical Analysis.

about the desk. Such analysis of common sense propositions about the desk leads to a set of basic premisses from which is synthesised a set of propositions which captures the truths embodied in the initial common sense propositions, yet avoids their shortcomings such as vagueness, ambiguity, etc. These *analysed results* are *substituted* for the initial *unanalysed* results about the desk. This completes the primary part of the Russellian analysis of the desk, i.e. the analysis of propositions.

The sense in which the desk as an *ontological object* is *analysed* is quite different. Russell's view is emphatically *not* one that has the physical desk of the metaphysics of common sense analysed into smaller parts and then resynthesised (say) as a swarm of sub-atomic particles (PLA: 161). (Though, of course, Russell claims all along that his theories are compatible with those of physics as realistically understood.) Instead ontological analysis is an *indirect* outcome of the analysis of propositions. Whereas analysis of propositions in the analysis diagram is represented by the two *vertical* arrows in the left hand half, analysis of ontological objects occurs via the *horizontal* inference arrows between the two halves of the diagram. The desk of the metaphysics of common sense is inferred uncritically from the initial set of common sense propositions. Since, as we have seen, a set of analysed results is *substituted* for the common sense propositions, and since, in addition, the desk of common sense cannot be logically inferred from the refined, substitute set of propositions, it follows that the desk of common sense has *no* ontological status, i.e. it doesn't exist.[2] If it doesn't exist, *a fortiori* it cannot be analysed into parts. Of course, this non-existence of the common sense desk doesn't falsify any *observations* made by the supporters of common sense. Rather, Russell will *substitute* something else for the desk, the substitute being equally compatible with any observations, but having tightly argued theoretical support of a kind that the common sense view lacks.

From whence does Russell obtain his ontological substitute for the common sense desk? As the analysis diagram shows, crucial ontological inferences occur at the level of *premisses*, i.e. faulty inference from unanalysed results is displaced by sounder inference from the premisses reached by the logical analysis step. (I say 'sounder' inference to emphasise the point that the premisses are unlikely to be final. Rather, as we have seen, they are liable to serve as the results from which further analysis will produce even more basic premisses.) It is the mark of premisses that they are precise and logically independent, i.e. they are propositions taking *basic* logical forms. It was Russell's hope that philosophical progress would result from determined efforts to evaluate in an open-minded manner the claims of the various logical forms when arriving at a set of premisses. In contrast to his predecessors who generally assumed that the subject-predicate form was basic, Russell proposed that various alternatives be tried. The clearest example of this is his work is the analysis of common sense propositions like 'I see a white patch' which were initially assigned the logical form 'This is white' but were subsequently included under the different form 'Whiteness is compresent with centrality'.[3] Since white is a *predicate* in the first of these forms and a *term of a relation* in the second, Russell was led to regard it as a *property* in the first case, but as a *particular* in the second one. (Though, as we will find in Part Two, he had other reasons for the rather odd claim that whiteness is a particular).

Such open-mindedness about the logical form appropriate to a given set of facts lies behind a well-known passage in the lectures on Logical Atomism. Having defined 'particulars' as "terms of relations in atomic facts", Russell adds:

> That is the definition of particulars, and I want to emphasise it because the definition of a particular is something purely logical. The question whether this or that is a particular, is a question to be decided in terms of that logical definition. In order to understand the definition it is not necessary to know beforehand 'This is a particular' or 'That is a particular'. It remains to be investigated what particulars you can find in the world, if any. The whole question of what particulars you actually find in the real world is a purely empirical one which does not interest the logician as such. The logician as such never gives instances, because it is one of the tests of a logical proposition that you need not know anything whatsoever about the real world in order to understand it.
>
> (PLA: 177).

This reinforces the present point being made about the analysis diagram: analysis is essentially applied to propositions and only indirectly to ontological objects. In our analysis of the desk Russell is urging us not even to start with the idea that because the common sense desk is a particular therefore in philosophical analysis its substitute will be a particular or group of particulars. Rather, we are asked to analyse common sense propositions about the desk into precise, logically independent premisses that are far from obvious in part, perhaps, because their logical form is markedly different from that of the common sense propositions. The justification for adopting a set of novel, unexpected premisses is simply that no other set appears to be as economical or efficient a starting point from which can be deduced the truths hinted at in a confused way by common sense knowledge.

All of this serves to emphasise why the first stage of analysis is the most *philosophical.* When Russell tells us that "philosophy, starting from data which are common knowledge, seeks to purify and generalise them into the simplest statements of abstract form that can be obtained from them by logical analysis" (*OKEW2*: 190),[4] he is not commending the practice of logic as an end in itself. Rather, it is by identification of appropriate logical forms that we can come to understand the basic categories of the world. As Chapters Six and Seven will demonstrate, Russell's analysis of empirical knowledge repeatedly centred on relations as a category that defied analysis into something else. In terms of our analysis diagram, the premisses frequently take the logical form of various relational propositions, so that it can be inferred that relations, in some sense, are part of the ultimate furniture of the world. This is the case, we will find, even though the premisses are typically amenable to yet further steps of analysis. This is so because such analysis tends to proliferate relation *tokens*, though not necessarily relation *types*. This trend amongst relations is not matched by properties and particulars, which prove more susceptible to replacement by something else as analysis is carried further. Thus, because our premisses might prove to be further analysable into even more basic premisses, inferences from such premisses to basic particulars, properties and relations are always somewhat provisional. Nevertheless the important point remains that, at any stage, our best estimate of what ontological entities are basic comes from basic premisses of analysis, rather than from any direct analysis of objects. If it should turn out that our current ontological entities are amenable to further analysis, Russell holds that this won't falsify any of the knowledge based on such entities.[5] Rather, the further analysis extends our knowledge with the only other change being that entities that were provisionally taken to be simple turn out to be complex.

The second important point about Russellian analysis suggested by the analysis diagram concerns the narrow and broad interpretations of analysis. We saw at the end of the previous chapter that there is a common tendency to adopt the narrow interpretation, i.e. to regard the first part of analysis (logical analysis on the diagram) as all there is to Russellian analysis. By contrast the broad interpretation regards both parts as essential to Russellian analysis (philosophical analysis on the diagram). What the diagram does is emphasise the real differences between the two interpretations at a metaphysical level. Consider again the analysis of the desk. For those who support the narrow interpretation of analysis, Russell started with common sense notions of a desk and by analysis reduced it to sense-data. The vague, complex, concrete everyday physical object has been "extruded from the world of what there is" (PLA: 237) and its place taken by a collection of more *simple*, transient particulars, viz., sense-data. On the narrow interpretation analysis finishes up at the simples (or at least what are then currently taken to be simples). However, for the broad interpretation analysis is not yet complete. The desk as a physical object is not replaced by sense-data *simpliciter*, but rather by a special compound arrangement of sense-data. The return journey of synthesis yields a *complex structure* of simples, i.e. a collection of sense-data having definite *relations* to one another. Thus, on the broad interpretation, analysis finishes at a *complex* composed of simples identified during the analysis. On the former view the simples alone matter; on the latter, the *relations* between the simples are equally as important as the simples themselves.

It is worth noting that the obsession of so many Russell commentators with ultimate simples is probably due to their adherence to the narrow interpretation.[6] If analysis finished at simples, it is a natural response to wonder whether rock bottom has been reached. On the other hand, if a major fruit of analysis is the capacity to construct complex structures whose interest is *unaffected by whether their simples are ultimate or not* (cf. quotes N and Q in Chapter Three) then clearly the question of ultimacy is not nearly so urgent. This was precisely Russell's view on the matter.[7] So then, the ambiguity that has been pointed out in the meaning of 'analysis', promotes confusion about whether analysis leads to simples or not. In the narrow interpretation it clearly does; in the broad one it does and it doesn't since it leads to complexes via simples. Sense-data are simpler than chairs in straightforward *epistemological* and *metaphysical* senses, whilst a logical construction of sense-data is more complex, *logically* speaking, than a chair. When Russell's own uses of 'simple' conflict, as in the cases noted above, it will be due to these reasons rather than any change of doctrine. Russell's own adherence to the broad interpretation of analysis is evident from some of the various quotations already given, e.g. "Out of the original manifold of events, logical structures can be built which will have properties. . . ." (J above). That Russellian analysis goes further than finding simples to replace complexes is demonstrated repeatedly in statements like the following:

> When some set of supposed entities has neat logical properties, it turns out, in a great many instances, that the supposed entities can be replaced by purely logical structures composed of entities which have not such neat properties.
>
> (LA: 164).[8]

However, to see fully why complex logical structures make such a difference to Russell's philosophy that the narrow interpretation invariably gets it hopelessly wrong, we need to consider the difficult topic of *relations*. This will be done in the next three chapters. In the meantime, the remainder of this chapter will consolidate the claim that the analysis diagram correctly represents Russell's conception of analysis at each stage of his career following his revolt against idealism. One outcome of this will be to locate the source of some common misconceptions about Russellian analysis.

4.1.2 Epistemological Propositions and the Analysis Diagram

It will be helpful to begin by examining the relation of *epistemological* considerations to the analysis diagram. The diagram as presented at the beginning of this chapter consists of a hierarchy of propositions in the left hand half and various inferred ontological items in the right hand half. Clearly, fundamental or primitive, epistemological propositions will figure amongst the left hand half hierarchy. Our present task is to decide exactly where they fit into this hierarchy of propositions.

For a start, we know that there is a strong epistemological motivation underlying Russell's major philosophising. Alan Wood, in a fragment specifically endorsed by Russell,[9] has claimed that Russell's "overriding motive was yearning for absolutely certain impersonal knowledge" (*MPD*: 192). Looking back on the occasion of his eightieth birthday, Russell himself wrote that up to the age of thirty eight he gave most of his energies to the task of finding out "whether anything could be known" because he "wanted certainty in the kind of way in which people want religious faith" (*PfM:* 53). As Wood saw it, Russell successively sought "to establish the truth of religion, the truth of mathematics and the truth of science", in the process developing "a distinctive philosophical method which added to knowledge, even though it could not confer certainty" (*MPD*: 194–5). The quest for certainty in mathematics and science was clearly a major thrust of Russell's mature philosophy and the failure of this quest had him writing by 1950 that the

> . . . demand for certainty is one which is natural to man, but is nevertheless an intellectual vice . . . What philosophy should dissipate is *certainty*, whether of knowledge or of ignorance . . . all our knowledge is, in a greater or less degree, uncertain and vague. . . .
>
> (PFL: 32–3).[10]

In the face of this early zeal for certainty, a clear answer is suggested to our question concerning the location of fundamental or primitive epistemological propositions in the left hand half of the analysis diagram. Following a long tradition, the premisses would be certain propositions thereby providing unshakable foundations[11] for the whole corpus of knowledge. Such a foundationalist view is common amongst interpreters of Russell.[12] As noted at the end of the previous chapter, Lakatos[13] views Russell's early logicism as belonging to "the *Programme of Trivialisation of Knowledge*" in which "all knowledge can be deduced from a finite set of trivially true propositions consisting only of terms with a trivial meaning-load" (1978b: 4–5). According to Lakatos, "the starting-point of Russell's philosophical career was to establish mathematics as the Euclidean bridgehead" (1978b: 11) from which a "later general attack" would seek "to Euclidise and trivialise the whole universe of knowledge". Thus Lakatos has the early Russell locating certainty in the initial premisses from which all else is deducible. On this view, analysis ceases when it arrives at the luminously obvious premisses from which all knowledge is derived.

Though common as an interpretation of Russellian analysis, the foundationalist view is nonetheless mistaken. It clashes with the characteristics of premisses listed in Table 3.1 (in the previous chapter) and with the analysis diagram (at the start of this chapter), both of which were derived from our wide selection of what Russell said about his method of philosophical analysis. However, so pervasive is the idea that Russell's philosophising is primarily epistemological and that certainty is thereby to be found, if at all, in the premisses, that more persuasive counter-arguments are needed.

These will be provided by a detailed consideration of the main phases of Russell's philosophical development in which it will be demonstrated that in each post-idealist phase, primitive epistemological propositions have a place in the analysis diagram, but not as the premisses reached by logical analysis of an advanced kind.

For present purposes, as well as the concerns of Part Two of this book, Russell's philosophy consists of his neo-Hegelian phase and then, following his revolt against idealism, falls into a further three distinct periods as follows:

– *A neo-Hegelian phase* (till 1899)
– *A Platonist phase* (1899 to 1913)
– *An Empiricist phase* (1914 to 1918), and
– *A Modified Empiricist phase* (from 1919 onwards).

Each of these will be discussed in turn.[14]

4.2 Russell's Neo-Hegelian Phase

4.2.1 Influences

As was noted at the end of the chapter on analysis in mathematical philosophy, Russell's early work in this area was especially influenced by Kant. In particular, his approach to philosophical analysis in *EFG* was seen to be based on typical Kantian patterns of argument. However, while the remaining work of Russell's idealist phase continues to employ this *method* derived in part from Kant, the work itself comprises a project that is distinctively Hegelian (albeit Hegel filtered through the ideas of various British idealists). This neo-Hegelian project is usually called the 'Tiergarten Programme' after the place in which Russell reported that he vividly conceived it.

> I remember a cold, bright day in early spring when I walked by myself in the Tiergarten, and made projects of future work. I thought that I would write one series of books on the philosophy of the sciences from pure mathematics to physiology, and another series of books on social questions. I hoped that the two series might ultimately meet in a synthesis at once scientific and practi-

cal. My scheme was largely inspired by Hegelian ideas.

(*Auto.*: 127).

While much of Russell's life's work can be seen as at least a partial fulfilment of the 'Tiergarten Programme', the immediate outcome of his neo-Hegelian philosophical efforts was an attempt on the series of books on the philosophy of the sciences.[15] In undertaking this project, Russell was influenced by the leading British idealists of the time, particularly Ward, McTaggart and Bradley. James Ward, Russell's most influential teacher at Cambridge, helped to shape his idealist thinking in a variety of ways.[16] A notable aspect of Ward's thought was his rejection of both 'top down' and 'bottom up' approaches to philosophising. While he was critical of the foundationalist approaches of the empiricists ('working from the bottom up'), he was also convinced that the Hegelian approach of starting with the grand metaphysical synthesis and working downwards was mistaken. His view was that "one had to begin in the middle, with actual scientific results" (Griffin, 1991: 40). Russell's proposal to analyse the various sciences and the dialectical relationships between them was an application of this advice. Russell's early idealist thinking was also partly influenced by the work of J.M.E. McTaggart, a contemporary at Cambridge, who was the leader of the pluralist strand of British idealism (F.H. Bradley, of course, led the monist strand). Despite some flirtations with monism, Russell was also a pluralist at this time (Griffin, 1991: 80). McTaggart's view that the Absolute was knowable was also in stark opposition to Bradley's position. McTaggart himself attempted an account of the Absolute by confronting it directly. Russell's very different approach was in accord with McTaggart's views in the sense that he hoped that his project to analyse the nature of the various sciences and the dialectical relationships between them would, if carried out in sufficient detail, reveal something of the nature of the Absolute. However, despite siding with McTaggart on pluralism and the knowability of the Absolute, the neo-Hegelian Russell was also heavily influenced by Bradley, particularly in matters to do with logic.

4.2.2 Russell's Neo-Hegelian Strategy

The nature and scope of Russell's neo-Hegelian philosophy has recently become much better understood due to the publication of his papers of the period (*Papers* 2), as well as Griffin's detailed study of this phase of his work, *Russell's Idealist Apprenticeship* (1991). In brief, Russell's project was to analyse the nature of the various sciences and the dialectical relationships between them. He was particularly concerned with the general question: Under what conditions are each of the special sciences possible? By the time that he came to abandon idealism, Russell's work on this overall project encompassed geometry, physics and pure mathematics. Whenever Russell analysed a science he encountered difficulties that his idealist presuppositions led him to view as contradictions. He saw these contradictions as the inevitable outcome of the limited and partial assumptions on which each special science rested. His hope was that the contradictions could be resolved by a dialectical transition through the sciences, e.g. the contradictions in geometry might be resolved in physics. Ultimately, of course, the final elimination of contradiction and paradox would require a dialectical transition or supercession to a wider, over-arching science whose special subject matter would be the Absolute.

As Griffin (1991: 316) points out, "Russell's neo-Hegelian paradoxes are essentially of two different kinds: . . . paradoxes of relativity and paradoxes of parts and wholes." Paradoxes of relativity were frequent in Russell's work from 1896 to 1899, ". . . although a common form for them did not emerge in Russell's writings until 1898" (Griffin, 1991: 317). According to Russell, "the contradiction of relativity" is "the contradiction of a difference between two terms, without a difference in the conceptions applicable to them" (AMR: 166). It occurs in mathematics and the sciences generally wherever there is a "conception of difference without a difference of conception" (RNQ: 81). An example is "the antinomy of the point" (EAE: 328). The concept of a point in geometry leads to contradiction because all points are intrinsically exactly alike (they have no "difference of conception"), yet each is distinct due to its relations to other points (hence there is a "conception of difference"). For neo-Hegelians, committed to the doctrine that different relations have to be grounded in different intrinsic qualities of their terms, a contradiction is thereby generated.

The paradoxes of parts and wholes were Russell's equivalents of the contradictions of infinity that had long been a staple of Hegelian orthodoxy. Such contradictions were supposed to arise in connection with infinitely divisible wholes ("paradoxes of infinite divisibility") or where discrete parts combine to form an infinite whole, such as points constituting space. As Griffin (1991: 316) points out, "[i]t is not hard to see connections between these paradoxes and some of the paradoxes which Russell was soon to make famous."

Finally, when the dialectical transition or supercession required for the successful completion of this neo-Hegelian project came to seem increasingly unlikely, Russell identified a common thread in these apparent contradictions: they each required a neo-Hegelian view of relations to generate the contradiction. The solution to the difficulties was to abandon neo-Hegelianism.[17] Russell's neo-Hegelian view of relations will be discussed in Chapters Five and Six.

4.2.3 Continuities and Discontinuities with Later Phases

Despite the abandonment of idealism for an extreme version of realism, quite a number of aspects of Russell's neo-Hegelian work were to become prominent features of his later work. These continuities include:

(a) Analysis as Philosophical Method. Russell's neo-Hegelian work clearly anticipates some of the major features of analysis discussed in earlier chapters, where it was found that Russellian analysis consists of movement from common knowledge (*results*) to abstract propositions (*premisses*), followed by the return journey to propositions which capture the truths of the common knowledge whilst avoiding the errors of the latter (*analysed results*). While Russell's method of analysis in his neo-Hegelian work differs from this in some important respects, it also shows some clear similarities, so that Russellian analysis might be said to have evolved from the earlier work.

In his neo-Hegelian phase Russell attempted to identify the *a priori* parts of each of the special sciences using a two-part Kantian transcendental deduction.[18] Here is Russell's account of this method:

> We may start from the existence of our science as a fact, and analyse the reasoning employed with a view to discovering the fundamental postulate on which its logical possibility depends: in this case, the postulate, and all which follows from it alone, will be *a priori*. *Or* we may accept the existence of the subject-matter of our science as our basis of fact, and deduce dogmatically whatever principles we can from the essential nature of this subject-matter. In this latter case, however, it is not the whole empirical nature of the subject-matter, as revealed by the subsequent researches of our science, which forms our ground; for if it were, the whole science would, of course, be *a priori*. Rather it is that element, in the subject-matter, which makes *possible* the branch of experience dealt with by the science in question.
>
> (APG: 291–2 = *EFG*: 5).

The method that starts from the science as a fact, and analyses its reasoning so as to discover its fundamental postulate(s) is called by Russell the 'analytic' or 'regressive' method; while the other method that dogmatically deduces principles from the essential nature of the subject-matter of the science is called the 'synthetic' or 'progressive' method. In his neo-Hegelian phase, "Russell applies the two methods together, so that the whole treatment is supposedly self-correcting" (Griffin, 1991: 82).

This neo-Hegelian approach to philosophical analysis clearly has strong similarities to the general account of Russellian analysis that has been developed in earlier chapters. Firstly, the 'analytic' or 'regressive' method is retained by Russell as the first part of his mature method of philosophical analysis, though the Kantian 'synthetic' or 'progressive' stage is rejected and replaced by something quite different. Secondly, the use of the two jointly self-correcting methods, the 'analytic' or 'regressive' method and the 'synthetic' or 'progressive' method parallels the later two-part view of analysis, backwards from the science as given (*results*) to abstract propositions (*premisses*), and then forwards from the premisses to a reconstruction of the original science (*analysed results*). Thirdly, the term 'analysis' is ambiguous. In the wider sense, it applies to the two-part process. In the narrower sense, it applies to the analytic method as against the synthetic one. Thus, when Russell joined Moore in the revolt against idealism, he took with him ideas about the nature of analysis that were to be transmuted into the distinctive approach that became Russellian analysis.[19]

(b) Reductionism. Inherent within Russellian analysis as described so far is the notion of reductionism. After all, the general scheme of the backwards movement from common knowledge (the *results*) to abstract propositions (the *premisses*), followed by the return journey to propositions which capture the truths of the common knowledge whilst avoiding the errors of the latter (the *analysed results*) already carries with it a hint of reductionism. Common knowledge can be regarded as reduced to the analysed results. Likewise if the results are mathematics and the premisses turn out to be logic, then mathematics is reduced to logic. There are obvious continuities here with the neo-Hegelian reduction of the special sciences into a science of the Absolute, and, more generally, of everything into the Absolute. The crucial difference, of course, is that the neo-Hegelian project represents an "upwards" reduction, or, more accurately, an augmentation. In the neo-Hegelian case, the results of the 'reduction' are not consistent until the Absolute is supplied. The analysis of geometry (or of the concept of space), e.g., doesn't result in a consistent theory. The initial results (e.g. the ordinary items of geometry), therefore depend for their truth upon the broader theoretical framework, including physics and, eventually, the theory of the Absolute.

(c) The Identification and Removal of Paradoxes/Contradictions. As already discussed, Russell's neo-Hegelian work featured various paradoxes/contradictions, which in standard fashion he attempted to resolve into a higher synthesis. Paradoxes/contradictions continued to play an important role in his post-idealist philosophising, the most famous, of course, being Russell's Paradox. His main concern was to show that the typical paradoxes/contradictions that Hegelians claimed to find in mathematics and science were really products of their own particular philosophical assumptions. On the other hand, any genuine paradoxes/contradictions in mathematics and science would need to be removed by appropriate changes within their basic premisses.[20]

(d) The Need to Develop an Alternative to Kant's View of Mathematics. Russell thought it unfortunate that the subjectivist doctrines of the *Critique of Pure Reason* had shaped idealist thinking on mathematics. Accordingly his *Essay on the Foundations of Geometry* set out to desubjectify Kant.[21] This aim continued to underpin his subsequent neo-Hegelian work on geometry and mathematics and carried over into his famous post-idealist work on the foundations of mathematics. The continuing theme in Russell's philosophy of the attempt to answer Kant will be discussed in Chapter Nine.

(e) Other continuities worthy of brief note include a commitment to pluralism, a cautious optimism about the possibility of attaining knowledge, and a belief in science as the best source of truth, hence the ongoing project to analyse the sciences. Another continuity, underpinning the project of analysing the sciences, was the belief that pure mathematics is concerned in an *a priori* way with *possible* spaces, Euclidean and non-Euclidean, and *possible* dynamics, Newtonian and non-Newtonian.

Major discontinuities between Russell's neo-Hegelianism and his later work include the change in the method of analysis discussed already, a vehement rejection of Kant (characteristic of Russell's writings from 1899 onwards as he came to view logicism as the answer to the question 'how is maths possible?'), the adoption of Peano's logic, and an unwavering commitment to external relations.

4.3 Russell's Platonist Phase

Labelling Russell's work from 1899 to 1913 as his 'Platonist phase' raises immediate questions such as: 'Why call it Platonist?' and 'Is 1899 to 1913 really a single phase, since Russell did change his mind on some important issues during that period?' In providing answers to these questions, the following discussion will be drawn inevitably into addressing Russell's theory of the logical subject and the concomitant method of analysis in the works from 1899 to 1913.

4.3.1 Why Call Russell's Work from 1899 to 1913 the 'Platonist Phase'?

The reason for calling this phase of Russell's work 'Platonist' is the standard one that throughout it Russell posits a densely populated world of abstract objects that have their being independently of knowing subjects. This period of Russell's work, represented for instance in *Principles of Mathematics* and *Problems of Philosophy*, is notable for its realm of "unchangeable, rigid, exact" objects that are "delightful to the mathematician, the logician, the builder of metaphysical systems, and all who love perfection more than life" (*PP*: 100). "Numbers, the Homeric gods, relations, chimeras and four-dimensional spaces" (*PoM*: 449) are some of the denizens of this world of *being*. "*Being* is that which belongs to every conceivable term, to every possible object of thought . . . " (*PoM*: 449). "*Existence*", on the other hand, "is the prerogative of some only amongst beings" (*PoM*: 449). Some existent objects are fundamental, in the sense that they are arrived at by *analysis* and survive further steps in analysis.[22]

So, in Russell's Platonist phase all objects are unchanging, independent and subsistent. Those objects that exist (i.e. non-abstract objects), are just special cases in this wider realm of objects. In opposition to the doctrines of idealism, Russell and Moore opted for a thoroughgoing anti-psychologism in which propositions (or judgements) are completely independent of any human acts of judging, analysing, etc. Thus propositions are viewed as independent, objective entities. For Russell, acquaintance with abstract objects like propositions is no more problematic than acquaintance with non-abstract objects. Hence, in this context, Platonism means much more than mere realism about universals. Of course, Russell maintained a realist theory of universals for the remainder of his career, thus in this weaker sense he remained a Platonist. However the reason for identifying an Empiricist phase starting in 1914 is precisely because Russell moved from the rich ontology of the Platonist phase to a much more austere substitute (as will be discussed in Section 4.4).

Russell himself, as late as *Problems of Philosophy*, in which his ontology had been pruned slightly, still called his work Platonist.[23] Amongst recent writers Griffin (1993a: 6) calls the underpinnings of *Principles of Mathematics* and beyond "radical Platonic realism". Hylton (1990: Part II) settles for "Platonic Atomism". Worries about calling this phase of Russell's work Platonist usually stem from historians of mathematics, who like to apply the term to theorists with an even richer ontology than Russell's. Thus Grattan-Guinness (1980: 88) describes Russell as "much less Platonist than Cantor in the assumption of abstract objects, especially after the discovery of the paradoxes." He also views Russell in *Principles of Mathematics* as having a "generally Aristotelian view" (1976: 160). However, as long as it is clear that 'Platonist' is being used in the accepted philosophical sense, there should be no confusion created.

Even so, it might be objected, are there not too many changes in Russell's thought from 1899 to 1913 for this period to be legitimately called a single phase? The answer is that from the ontological perspective, which is what the label 'Platonism' refers to, there weren't many changes at all. However, as shown by the (until recently) unpublished papers for the years 1898 onwards, Russell tried many theoretical changes during this period as he sought to develop a theory of terms that would be free of inconsistencies. The result is that what minor changes there were in ontology were not motivated by a desire for ontological reduction, but rather by the need to remove contradictions/paradoxes in the theory of terms which centred on Russell's notion of the logical subject. As Griffin (1993a) shows, the contradictions/paradoxes in Russell's evolving theory of terms were of two related kinds, which recurred repeatedly. Griffin calls them 'problems of complexes' and 'double aspect problems'. Over about a ten year period Russell's papers, both published and unpublished, constituted a kind of 'laboratory of experimental ideas' as he tried to perfect his theory of terms. According to Grattan-Guinness (1977: 156) this is ". . . one reason why *Principia Mathematica* is in some respects a rather confused work; Russell had tried so many alternatives during the fallow years that he could not always assess them as relevant or irrelevant to the final system adopted."

Against this background of continually refining the theory of terms so as to free it from inconsistencies, two key theses about Russell's philosophy from 1899–1913 warrant it being called the 'Platonist phase'.

1. There were no really major changes in ontology, except that denoting concepts (i.e. the meanings of denoting phrases), propositions and classes were eliminated. All of these eliminations were a result of moves to resolve contradictions/paradoxes in the theory of terms.[24]
2. The theory of descriptions was not mainly, or at all, about ontological reduction; it was about eliminating *denoting concepts*. The gist of 'On Denoting' (1905) is that denoting concepts are incoherent.

Evidence for the truth of these theses is presented in the next section and in section 7.2

4.3.2 Russell's Theory of the Logical Subject

The strength of the claim that there were no really major changes in ontology during the Platonist phase can be judged from the fact that continually throughout 1899–1913 Russell's ontology included: universals, sense-data, material bodies, minds, points, and instants. According to Russell, only the first two and (possibly) one's own mind are known by acquaintance – the others being known by description.[25]

The ontological reduction that did occur in the Platonist phase can be broadly[26] summarised as follows:

– 1901: Russell's logicism reduces numbers and mathematical entities to sets of sets.[27]

– 1905: Denoting concepts are dispensed with via the theory of descriptions.

– 1906–8: Attempts are made to dispense with classes and propositions via the substitutional theory.

– 1910: Propositions are reduced via the multiple relation theory.[28]

– *PM*: Classes are reduced to propositional functions and universals.

As Cocchiarella (1987: 7) points out, during the Platonist phase the "... fundamental notion ... that provides the key toward understanding Russell's changing position ... is the notion of a logical subject, i.e. the notion of an entity that can be *named*, *mentioned*, or *denoted*, and about which something can be asserted." In *Principles of Mathematics*, logical subjects (or terms) include properties, relations and propositions. All have being in the same sense that concrete particulars do; i.e. all properties, relations and propositions are individuals, as are concrete particulars, and all are values of individual variables. This world of being extends to mathematics:

> Arithmetic must be discovered in just the same sense in which Columbus discovered the West Indies, and we no more create numbers than he created the Indians.
>
> (*PoM*: 451).[29]

Thus, for Russell 'term' is "the widest word in the philosophical vocabulary" (*PoM*, #47). Reflecting Peano's influence, logic becomes a calculus of *terms*. This results in a sharp contrast with traditional logic and the subject-predicate doctrine that underpins it. Russell denies that there is a distinction in kind between subjects and predicates, treating both alike as terms. This has important metaphysical implications:

> The case which, on the old theory, would be treated as the inherence of a quality *q* in a substance *s*, is to be treated on the new theory as the combination of two terms *q* and *s* into a complex term *q-s*. The scholastic distinction between substance and accident was thereby obliterated; items in both scholastic categories are terms.
>
> (Griffin, 1992: 195–6).

One notable feature of this calculus of logical subjects or terms is that, as well as the categories of truth and falsehood, it includes that of meaninglessness. Although *Principles of Mathematics* is not much concerned with linguistic matters, Russell appears to have added the references to meaninglessness late in the writing, so as to provide an initial way of dealing with the Russell paradox (*PoM*, Appendix B).[30] Another notable feature of this theory of terms is that it is the basis of the theory of denoting, which, with its introduction of the notion of a *denoting concept*, is one of the main innovations in *Principles of Mathematics*.[31] As *Principles of Mathematics* is usually interpreted, certain words or phrases indicate concepts which denote terms. Such concepts are called denoting concepts and the crucial point is that denoting is a relation between concepts and terms, not between words and terms. Given this, the common interpretation is that every denoting concept does in fact denote. Hence Russell is credited with a bloated ontology that includes all kinds of non-existent terms, such as chimaeras, very much like Meinongian non-existent objects.

In fact, as Cocchiarella (1987, Ch. 3) has shown, Russell's pre-1905 ontology may not have been as lush as has been generally assumed. According to Cocchiarella, from the start of Russell's post-idealist philosophy not all denoting concepts denote, instead many of them merely purport to denote. Hence, on Cocchiarella's interpretation, Russell's pre-1905 ontology does not include a plethora of non-existent terms. 'Chimaera', e.g., is a denoting concept that fails to denote. Evidence that the Cocchiarella interpretation is correct is provided by Griffin (1993b). The oft-quoted passage from *Principles of Mathematics* that is usually taken to underpin the common interpretation is:

> Whatever may be an object of thought, or may occur in any true or false proposition, or can be counted as *one*, I call a *term* . . . every term has being, *i.e. is* in some sense. A man, a moment, a number, a class, a relation, a chimaera, or anything else that can be mentioned, is sure to be a term; and to deny that such and such a thing is a term must always be false.
>
> (*PoM*: 43).

The "object of thought" at the beginning of this quotation is usually taken to be a *term* that is denoted by a denoting concept. However, as Griffin shows, in Russell's most explicit outline of the semantics of *Principles of Mathematics* (in a 1904 letter to Victoria Welby), his intention is absolutely clear that the "object of thought" is a *denoting concept* (1993b).[32] Griffin goes on to consider whether any passages of *Principles of Mathematics* conflict with the Cocchiarella interpretation. His conclusion is that while no passage clearly contradicts it, it has the merit of solving a variety of difficulties for the common interpretation, such as Russell's account of the null class.

Overall then, there is a strong case for maintaining that Russell's pre-1905 ontology was not as lush as is usually supposed. As Cocchiarella points out, shorn of the plethora of non-existent objects, but still including denoting concepts, the logic of *Principles of Mathematics* is similar in many respects to recent intensional logics developed by analytical philosophers. "Ironically, these logics are even based for the most part on essentially the same sort of realist metaphysical framework that Russell had advocated in his early philosophy" (Cocchiarella, 1987: 5).[33]

The Cocchiarella interpretation also challenges the standard view of the theory of descriptions. According to the standard view, the theory of definite descriptions, and its subsequent generalisation into the theory of incomplete symbols, enabled Russell to drastically lop his previous bloated Meinong-like ontology of *Principles of Mathematics*. A more accurate view, stemming from the Cocchiarella interpretation, is that the theory of descriptions showed how to get rid of denoting concepts and thereby removed a major source of inconsistency in the theory of terms (Griffin, 1993a: 17ff.). Thus Russell introduced denoting concepts in *Principles of Mathematics* as part of his continual refining of the Platonist phase theory of terms so as to free it from inconsistencies. In the event, denoting concepts created as many difficulties as they solved and were made redundant by the theory of descriptions, as Russell sought a way of referring to the meaning of a descriptive phrase without creating contradictions. However, it should be noted that the theory of descriptions did not reduce the importance that Russell attached to the 'logical subject'. Rather the focus now shifted to the logical subjects of fully analysed sentences. The ontological implications of the theory of descriptions will be discussed further in Section 7.2.

While the elimination of denoting concepts is a significant ontological reduction, it is much more modest than the ontological reduction claimed by the standard view. Given that throughout 1899–1913 Russell's ontology continually included universals, sense-data, material bodies, minds, points, and instants, those changes in ontology that have been outlined in this section can be seen to be rather minor. This adds weight to the claim that this whole period is appropriately labelled 'Platonist'.

4.3.3 Misunderstandings of Russellian Analysis in the Platonist Phase

An excellent example of misunderstanding Russell's work in his Platonist phase is provided by Lakatos, who saw him as attempting to trivialise mathematics by deducing it from manifestly true premisses. Lakatos' claim seems to be partly inspired by a belief that a Platonist account of mathematical entities entails absolutely self-evident premises (see, e.g., 1978b: 18, 25). Why Lakatos believed this I don't know. He never argues for the view, but just takes it that that is Russell's position. On my reading of the two Russellian sources on Platonism in mathematics referred to by Lakatos (SOM and *MPD*, Ch. 17), there is no suggestion that the premisses are trivial or self-evident. What I *do* find is intellectual wonder at mathematics possessing "supreme beauty – a beauty cold and austere, like that of sculpture" and "the sense of being more than man, which is the touchstone of the highest excellence. . . . " (SOM: 86).[34] It is the "characteristic excellence of mathematics" that "the reasoning is rigidly logical", endeavouring "to present whatever is most general in its purity, without any irrelevant trappings" (SOM: 87). None of this implies anything about the nature of the premisses of logical analysis. True, "the central principles of logic" (SOM: 89) are numbered amongst the premisses, but these "principles . . . are so embedded in our ratiocinative instincts, that they are employed quite unconsciously, and can be dragged to light only by much patient effort" (SOM: 90). Once again, none of this implies that these "fundamental laws" (SOM: 90) are self-evident or trivial. As we will find shortly, Russell *does*, as a matter of fact, hold that once the basic principles of deduction are discovered, they *do* appear to be indubitable. But this does *not* apply to the *rest* of the premisses and is, in any case, independent of his Platonism.

So Lakatos' conviction that Platonist premisses are trivial appears to derive from his own preconceptions rather than from Russell. As Lakatos tells it, shortly after the *Principles of Mathematics*, Russell was confronted by difficulties in the Euclidean programme, difficulties that multiplied over the years so that Russell's subsequent development "is virtually the piecemeal retreat from Euclideanism, bravely fighting every inch of the way, and rescuing as much certainty as he could" (1978b: 11).[35] In the process, according to Lakatos, Russell adopts a more *inductive* view of mathematics and logic (1978b: 16–7, 25). For Lakatos the change is nicely captured by the contrast between Russell's 1901 claim for mathematics that "its edifice of truths stands unshakable and inexpugnable to all the weapons of doubting cynicism" (SOM: 93)[36] and the 1959 admission that "the splendid certainty which I had always hoped to find in mathematics was lost in a bewildering maze" (*MPD*: 157). One can agree with Lakatos that this loss of certainty is due to the need to include far from satisfactory components such as the axioms of reducibility, infinity and choice and ramified type theory. However, once again, none of this entails that the original certainty resided in trivial, self-evident premisses.

Anyway, Lakatos *ought* to have had doubts about his understanding of Russell's mathematical philosophy if only because just before the 1959 passage just noted (that Lakatos twice quotes), Russell has this to say about his abandonment of Platonism:

> Mathematics has ceased to seem to me non-human in its subject-matter. I have come to believe, though very reluctantly, that it consists of tautologies. I fear that, to a mind of sufficient intellectual power, the whole of mathematics would appear trivial, as trivial as the statement that a four-footed animal is an animal.
>
> (*MPD*: 157).

Here we seem to have Russell claiming that his development is somewhat contrary to what Lakatos believes, i.e. instead of starting with a trivialised version of mathematics and moving to an inductive version having substantial content, Russell sees it happening more the other way around. However, as we found in Chapter Two, this doesn't mean that Russell came to believe that analysis stopped at tautologous premisses. Rather, there is still scope to analyse the *concepts* that are taken as primitive at the level of analysis involving the tautologous premisses.

However, the really decisive point against the Lakatos account of Russell's early mathematical philosophy is that Russell himself expressly contradicts its characterisation of the premisses. According to Lakatos, Russell's Euclidean programme was developed at least up to *Principles of Mathematics*, after which difficulties gradually set in. Yet right at the start of his Platonist phase we find Russell taking Leibniz to task precisely because his 'Universal Mathematics' had analytic, i.e. trivial, premisses. Rather, says Russell, in regard to the premisses the "emphasis should be laid on the indefinable and indemonstrable. . . . " (*PL*: 170–1). In *Principles of Mathematics*, the "indefinables" are the "logical constants" and the "indemonstrables" are the premisses "concerned exclusively with logical constants and with variables" (*PoM*: 8). In this work, the premisses of mathematics belong to logic precisely because the indefinables are logical notions such as implication, relation, propositional function, class, etc.[37] "Formally" says Russell, "my premisses are simply assumed; but the fact that they allow mathematics to be true . . . is surely a powerful argument in their favour" (*PoM*, 1902 Preface: xviii). No hint here of Lakatos' trivialisation of mathematics! Certainly Russell was under no illusions about people finding his indefinables self-evident:

> The discussion of indefinables – which forms the chief part of philosophical logic – is the endeavour to see clearly, and to make others see clearly, the entities concerned, in order that the mind may have that kind of acquaintance with them which it has with redness or the taste of a pineapple. Where, as in the present case, the indefinables are obtained primarily as the necessary residue in a process of analysis, it is often easier to know that there must be such entities than actually to perceive them; there is a process analogous to that which resulted in the discovery of Neptune, with the difference that the final stage – the search with a mental telescope for the entity which has been inferred – is often the most difficult part of the undertaking.
>
> (*PoM*, 1902 Preface: xv).

I conclude then that, *contra* Lakatos, Russell in his early Platonist phase did *not* set out to deduce mathematics from indubitable premisses, but rather from premisses of the opposite sort as suggested in Table 3.1 and the analysis diagram.[38] It remains to show that this was so for the whole of the Platonist phase. In the process, the location of primitive epistemological propositions in the left hand half of the analysis diagram will be ascertained.

4.3.4 Russell on His Method of Analysis in the Platonist Phase

Probably the best statement by Russell about his philosophical method in the Platonist phase is in the 1907 paper "The Regressive Method of Discovering the Premises of Mathematics". Lackey has written of the views presented in this paper that they "are remarkable and have been little noted",[39] but they fit the above account of Russellian analysis like a glove. The objects of the paper are:

> . . . to explain in what sense a comparatively obscure and difficult proposition may be said to be a premise for a comparatively obvious proposition, to consider how premises in this sense may be discovered, and to emphasise the close analogy between the methods of pure mathematics and the methods of the sciences of observation.
>
> (RMDP: 272).

Regarding the word 'premise', Russell says that it has two quite different senses:

> . . . there is what we may call the 'empirical premise', which is the proposition or propositions from which we actually are led to believe the proposition in question; and there is what we will call the 'logical premise', which

is some logically simpler proposition or propositions from which, by a valid deduction, the proposition in question can be obtained. Thus, the empirical premises for 2 + 2 = 4 will be '2 sheep + 2 sheep = 4 sheep', and other like facts; whereas the logical premises will be certain principles of symbolic logic. (RMDP: 272–3).

What Russell here calls the 'empirical premise' represents the obvious unanalysed results for a primordial stage of analysis which had the then far from obvious premise 2 + 2 = 4. Today we find 2 + 2 = 4 obvious and it serves as the unanalysed results for a further stage of analysis whose logical premisses are various principles of symbolic logic. The "order of knowledge" (RMDP: 282) is the "proposition '2 sheep + 2 sheep = 4 sheep' was probably known to shepherds thousands of years before 2 + 2 = 4 was first discovered. . . . " (RMDP: 272). Likewise, the latter was known for a very long time before the recent discovery of its logical premisses. So at each stage of analysis the premisses are less obvious than the unanalysed results. For us, 2 + 2 = 4 is what I have been calling a primitive epistemological proposition. Clearly such propositions are not the premisses of our analysis diagram, but rather belong in the unanalysed results.

Since "the inferring of premises from consequences is the essence of induction" it follows that "the method in investigating the principles of mathematics is really an inductive method, and is substantially the same as the method of discovering general laws in any other science" (RMDP: 274). Thus, continues Russell, in "every science, we start with a body of propositions of which we feel fairly sure . . . commonly called the facts, which are generally got by observation". These unanalysed results are said to be more obvious, easier to grasp and logically more complex in relation to their premisses which are more obscure, harder to grasp and logically simpler. Since "premises which are ultimate in one investigation may cease to be so in another; that is, we may find logically simpler propositions from which they may be deduced" (RMDP: 273), it follows that *the more developed a science the more removed are its current premisses from our primitive epistemological propositions.* This results in what Russell calls an "apparent absurdity". Why proceed

> . . . through many rather recondite propositions of symbolic logic, to the 'proof' of such truisms as 2 + 2 = 4: for it is plain that the conclusion is more certain that the premises, and the supposed proof therefore seems futile. But of course what we are really proving is not the truth of 2 + 2 = 4, but the fact that from our premisses this truth can be deduced.
> (RMDP: 272).

What advantages are there in this? Russell gives three virtues in reaching even simpler logical premisses:

(i) . . . it gives a greater chance of isolating a possible pervading element of falsehood . . .
(ii) . . . it organises our knowledge . . .
(iii) . . . The logical premises have, as a rule, more consequences . . . and thus

lead to the discovery of many things which could not otherwise be known. (RMDP: 275).

There is much more on these and similar themes in the Russell paper under consideration,[40] however, we have seen enough, I think, to establish firmly the validity of the analysis diagram for the Platonist phase. Primitive epistemological propositions for reasonably advanced mathematics or science are clearly located in the unanalysed results (and *not* the premisses). As against foundationalist accounts of Russellian analysis which hold that logic justifies mathematics, it might be more correct to say that mathematics justifies logic, since mathematics is mostly more certain than its logical premisses. I say 'mostly more certain' here because Russell does, in fact, admit that strictly speaking *some* of the premisses are just as certain as the results deduced from them. This is so because the deduction of the more obvious results from the less obvious premisses depends on rules of deduction concerning many of which "it is almost impossible to entertain any reasonable doubt" (RMDP: 281).[41] Hence when 'premisses' is understood in this strict sense, some of the premisses are likely to be more obvious than some of the results. However, none of this affects the validity of our analysis diagram, as the following figure shows:

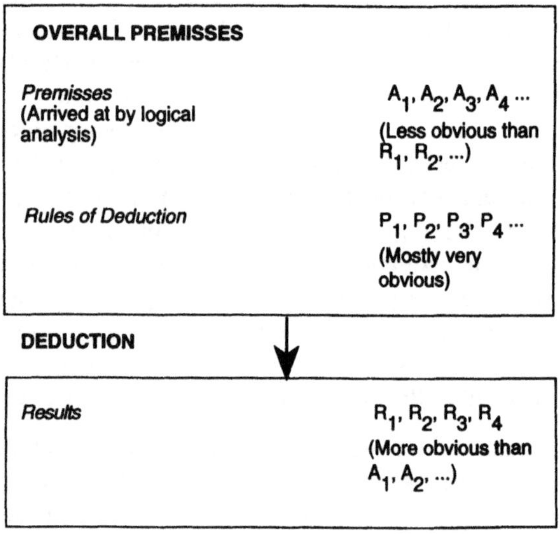

Figure 4.2.

Russell illustrates the point with a reconstruction of the Fregean analysis of mathematics involving three kinds of premisses:

a) some "indubitable" principles of deduction;

b) some "nearly as indubitable" principles of quantification; and

c) some "less evident, but indispensable" premisses proper.

While principles of deduction, such as "if either p is true or p is true, then p is true" (RMDP: 281), are even more obvious than the mathematics to be deduced, the premisses proper, such as the axiom of infinity are less obvious, in accordance with our analysis diagram.

Not surprisingly, the details emerging from our discussion of the 1907 paper illuminate and complement various other utterances of the Platonist phase, e.g. in the light our claim that it is more correct to say that mathematics justifies logic, the following passage from *Principia Mathematica* is pertinent.

> . . . (I)n fact self-evidence is never more than a part of the reason for accepting an axiom, and is never indispensable. The reason for accepting an axiom, as for accepting any other proposition, is always largely inductive, namely that many propositions which are nearly indubitable can be deduced from it, and that no equally plausible way is known by which these propositions could be true if the axiom were false, and nothing which is probably false can be deduced from it. If the axiom is apparently self-evident, that only means, practically, that it is nearly indubitable; for things have been thought to be self-evident and have yet turned out to be false. And if the axiom itself is nearly indubitable, that merely adds to the inductive evidence derived from the fact that its consequences are nearly indubitable: it does not provide new evidence of a radically different kind. Infallibility is never attainable, and therefore some element of doubt should always attach to every axiom and to all its consequences. In formal logic, the element of doubt is less than in most sciences, but is not absent, as appears from the fact that the paradoxes followed from premisses which were not previously known to require limitations. In the case of the axiom of reducibility, the inductive evidence in its favour is very strong, since the reasonings which it permits and the results to which it leads are all such as appear valid.
>
> (*PM*, Vol. I: 59).[42]

Similarly, the preceding discussion clarifies an apparent inconsistency in the 1911 paper "The Philosophical Importance of Mathematical Logic". In some comments on knowledge that is "primitive and undemonstrated", i.e. our primitive epistemological propositions, Russell says that it is

> . . . divided into two kinds: knowledge of particular facts, which alone allows us to affirm existence, and knowledge of logical truth, which alone allows us to reason about data.
>
> (PIML: 38).

Concerning the latter, Russell adds:

> In order that such a knowledge be possible, it is necessary that there should be *self-evident logical truths*, that is to say, *truths which are known without demonstration. These are the truths which are the premises of pure mathematics* as well as of the deductive elements in every demonstration on any subject whatever.
>
> (PIML: 39, my italics).

At first sight Russell might seem to have now adopted the view that was firmly repudiated above, viz. that the logical premisses of mathematics are purely self-evident and indubitable. That this is *not* the case, but rather a reference to the basic principles of deduction, which are near enough to indubitable, but only *some* of the premisses of mathematics, is clear from what he says on the very next page:

> On the subject of self-evident truths it is necessary to avoid a misunderstanding. Self-evidence is a psychological property and is therefore subjective and variable. It is essential to knowledge, since all knowledge must be either self-evident or deduced from self-evident knowledge. But the order of knowledge which is obtained by starting from what is self-evident is not the same thing as the order of logical deduction, and we must not suppose that when we give such and such premises for a deductive system, we are of the opinion that these premises constitute what is self-evident in the system. In the first place self-evidence has degrees: It is quite possible that the consequences are more evident than the premises. In the second place it may happen that we are certain of the truth of many of the consequences, but that the premises only appear probable, and that their probability is due to the fact that true consequences flow from them. In such a case, what we can be certain of is that the premises imply all the true consequences that it was wished to place in the deductive system. This remark has an application to *the foundation of mathematics, since many of the ultimate premises are intrinsically less evident than many of the consequences which are deduced from them.* Besides, if we lay too much stress on the self-evidence of the premises of a deductive system, we may be led to mistake the part played by intuition (not spatial but logical) in mathematics. The question of the part of logical intuition is a psychological question and it is not necessary, when constructing a deductive system, to have an opinion on it.
>
> (PIML: 40, my italics).

The 1907 paper, which we took to exemplify Russell's views on philosophical method in the Platonist phase, deals with mathematical philosophy. It is, therefore, reassuring to find that the 1912 *Problems of Philosophy,* his first major venture into general philosophy, conforms to the pattern already established. In particular, the sharp distinction that has been drawn between primitive epistemological propositions and the premisses reached by logical analysis is clearly marked in this work which is more concerned with theory of knowledge than with metaphysics (*PP*: 6). The book starts with the question of whether there is "any knowledge in the world which is so certain that no reasonable man could doubt it" (*PP*: 7). Before long Descartes' method of systematic doubt, and its outcome "that subjective things are the most certain" (*PP*: 18), is being commended, though Russell concludes that 'I think, therefore I am' says "rather more than is strictly certain." This is so because the self that does the thinking is almost as experientially elusive as the real table popularly supposed to lie behind the sense-data. Thus Russell concludes that our sense-data have "primitive certainty" i.e. they provide "a solid basis from which to *begin* our pursuit of knowledge" (*PP*: 19, my italics). So in general philosophy, self-evident statements about what is given in sense are primitive epistemological propositions. Like their counterparts in mathematical philosophy, e.g. 2 + 2 = 4, they are part of the unanalysed results, rather than the premisses, for any advanced stage of analysis.

In the course of *Problems of Philosophy*, Russell arrives at a scale of self-evidence which locates different sorts of propositions at different levels. The propositions with the highest degree of self-evidence (what Russell calls *intuitive* knowledge) include "those which merely state what is given in sense, and also certain abstract logical and arithmetical principles" (*PP*: 109).[43] So intuitive knowledge includes basic principles of deduction, which we found are part of the premisses of any deductive system, as well *as* both kinds of primitive epistemological proposition which are part of the unanalysed results for mathematical or scientific philosophy. Just below the highest degree of self-evidence come propositions about immediate memory, with the inductive principle, remoter memories and complicated parts of logic and mathematics figuring further down the scale (*PP*: 117). It will turn out to be a characteristic feature of Russell's philosophy that whenever he commends Descartes' method of systematic doubt, he is, in effect, constructing a scale of degrees of self-evidence of propositions as a preliminary to analysis, not as might be thought, arriving at the premisses of analysis. So, for instance, the discussion assigning "primitive certainty" to our sense-data is described as "the attempt to get clear as to our *data* in the way of knowledge of existence" (*PP*: 60, my italics).

Though *Problems of Philosophy* concentrates on epistemology it inevitably raises ontological questions. Physical objects and other people's minds are definitely ruled out as objects known by acquaintance, yet Russell wishes to retain them as objects of knowledge. The answer, of course, is that we have *knowledge by description* of such objects, i.e. knowledge by description "enables us to pass beyond the limits of our private experience" (*PP*: 59). Analysis thereby will lead to premisses going beyond mere objects of acquaintance. The significance of this becomes clear from a consideration of the different analysis diagrams that result from a foundationalist interpretation of *Problems of Philosophy*, as against the interpretation that we have been urging. The foundationalist interpretation entails the following analysis diagram (shown in Figure 4.3 and hereafter called the *foundationalist diagram*).

The foundationalist diagram restricts the premisses to what Russell calls *intuitive knowledge*, i.e. (as we found above) propositions with the highest degree of self-evidence. If intuitive knowledge constitutes immediate knowledge of truths, its counterpart in the ontological order is *acquaintance* which provides immediate knowledge of *things*. The things with which we are immediately acquainted are sense-data, ourselves (probably) and universals (*PP*: 109). However Russell's rejection of solipsism shows that the premisses in the foundationalist diagram are *not* his premisses since they are precisely the premisses for the solipsist position that "the world consists of myself and my thoughts and feelings and sensations, and that everything else is mere fancy" (*PP*: 22). While admitting that there "is no logical impossibility" in this position, Russell nevertheless rejects it on the grounds that

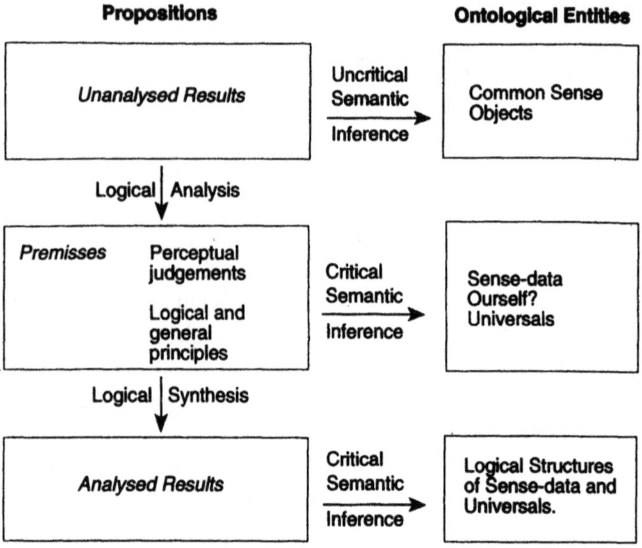

Figure 4.3. The foundationalist analysis diagram.

> . . . it is . . . a less simple hypothesis, viewed as a means of accounting for the facts of our own life, than the common-sense hypothesis that there really are objects independent of us, whose action on us causes our sensations
>
> (*PP*: 22–3).

Besides the material objects that cause our sensations, the Russell of the Platonist phase also admits as objects independent of us a host of other entities. Hence the following analysis diagram (hereafter called the *non-foundationalist*

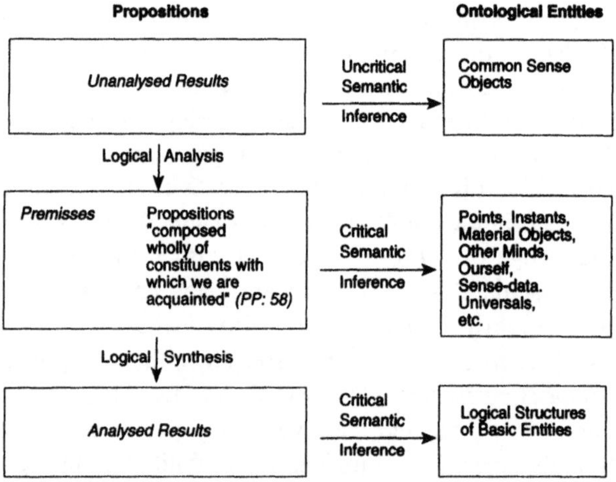

Figure 4.4. The non-foundationalist analysis diagram.

diagram), better represents Russell's position in *Problems of Philosophy*. According to the non-foundationalist diagram, basic entities include such things as points and instants which lie outside of experience. It is Russell's position that "there is no reason why I should not know of the existence of something with which *nobody* is acquainted" (*PP*: 45). But doesn't this contradict Russell's "fundamental principle in the analysis of propositions", the principle of acquaintance which runs as follows: *"Every proposition which we can understand must be composed wholly of constituents with which we are acquainted"*? (*PP*: 58). Not at all. Our acquaintance with universals is responsible for our *concepts* "such as *whiteness*, *diversity*, *brotherhood*, and so on" (*PP*: 52). Such concepts presumably include *material objecthood*, *pointhood*, *instanthood*, etc. since Russell notes that for

> . . . knowledge by description concerning things of which we have no direct experience . . . we need some connexion of universals, enabling us, from such and such a datum, to infer an object of a certain sort as implied by our datum. Thus in regard to physical objects, for example, the principle that sense-data are signs of physical objects is itself a *connexion of universals*; and it is only in virtue of this principle that experience enables us to acquire knowledge concerning physical objects.
>
> (*PP*: 148–9, my italics).

Hence, just as "we can understand propositions about 'the author of Waverley', without knowing who he was" (KAKD: 159), we can understand propositions about entities not directly known by anyone. Indeed in the Platonist phase it is the non-experienced entities such as points and instants, rather than experienced entities such as sense-data, which are fundamental in analysis. This point will be pursued further in our detailed consideration of the Platonist phase in Chapter Ten.

In this chapter we have been attacking the view that the premisses reached by Russellian analysis are primitive epistemological propositions. To this end we have shown that in major writings of the Platonist phase, Russell views primitive epistemological propositions as *data* on which to begin analysis. Typically, as analysis is further advanced, it reaches *premisses* which are *not* basic epistemological propositions. The only exceptions to this trend are self-evident principles of deduction which can be said to constitute a *part* of the premisses. Hence Russellian analysis in the Platonist phase complies with Table 3.1 in the previous chapter and the analysis diagram at the start of this chapter.

4.4 Russell's Empiricist Phase

Russell's Empiricist phase, exemplified in such works as *Our Knowledge of the External World* and "The Philosophy of Logical Atomism", is marked by a retreat from the realm of subsistent being characteristic of the Platonist phase. Developments such as the theory of descriptions had stimulated a gradual pruning of the world of being. However the crucial catalyst for Russell's move to his Empiricist phase was Whitehead's demonstration that points and instants could be logically constructed from sense-data.[44] Such logical constructions appear in his writings from 1914. Since points and instants had been for Russell paradigmatic cases of Platonic existents hidden from the world of sense, this development undercut the Platonism and led to a new emphasis on sense-data and sensibilia. Here, if ever, it might be anticipated that Russell's premisses in the analysis diagram would coincide with primitive epistemological propositions. After all, their divergence in the Platonist phase was only to be expected given that doctrine's characteristic postulation of a gulf between appearance and reality, but now that entities are to be constructed out of sense-data there seems to be every chance that the basic items of epistemology and ontology will be the same.

Certainly this is a common interpretation of Russell's Empiricist phase,[45] perhaps best summed up in a carefully argued paper by W. Lycan (1981). In this paper Lycan seeks to elucidate the arguments in favour of what he takes to be "the core thesis" of Russell's logical atomism, viz. "that all and only logical atoms are *ontological* atoms." His initial strategy is to test the theory that it is the epistemological properties of Russellian sense-data that form the bridge between logical and ontological atoms. Accordingly he first seeks to establish the two biconditionals:

(i) All and only logical atoms are epistemological atoms; and
(ii) All and only epistemological atoms are ontological atoms.[46]

The truth of these biconditionals would entail an analysis diagram in which (a) the limit of analysis had been reached; and (b) the premisses were primitive epistemological propositions containing logically proper names. This would imply ontological entities in the middle right hand half of the diagram that are simultaneously logical atoms, epistemological atoms and ontological atoms. Clearly this is at odds with our account of the analysis diagram, already established for the Platonist phase, in which primitive epistemological propositions are *not* the key premisses of any advanced stage of Russellian analysis.

Let us examine in turn Lycan's attempts to establish the two biconditionals. In seeking to establish that: (i) all and only logical atoms are epistemological atoms, Lycan finds it helpful to work with the contrapositives of the two halves of this first biconditional, i.e.

–(EL) All epistemological fictions are logical fictions, and
–(LE) All logical fictions are epistemological fictions.

In support of (EL), Lycan cites two passages from Lecture VIII of "The Philosophy of Logical Atomism". One passage concerns our experience of physical objects (PLA: 237), whilst the other does the same for persons (PLA: 239–40).[47] As Lycan sees it, in both passages "Russell argues explicitly from something's being an epistemological fiction to that thing's being a logical fiction" (Lycan: 211). On the contrary, I will argue that while Russell does hold that physical objects are both epistemological and logical fictions, in the case of persons he views them as epistemological fictions but denies they are logical fictions in Lycan's sense. So if my argument is correct, (EL) is false.

In the Platonist phase Russell had flirted with the notion of direct acquaintance with the self without being fully convinced.[48] This tentative view was replaced at the start of the Empiricist phase by the doctrine that while "subjects are not given in acquaintance", they "are known . . . as referents for the relation of acquaintance" (ONA: 164). So, when

> . . . two objects O and O' are given as parts of one experience, we perceive the fact 'something is acquainted with both O and O''. Thus two instances of acquaintance can be given as having a common subject, even when the subject is not given.
>
> (ONA: 165).

From this and similar considerations, Russell concludes that by "the help of reflection and special experiences" (ONA: 168) we arrive at the self or subject as an entity. This conclusion is, of course, connected with Russell's well known rejection of neutral monism in the Empiricist phase.[49] As against the neutral monists, Russell maintained that sensation and perception, for example, are irreducible two-term relations between a subject and particulars and facts respectively. In this period Russell had *no analysis* of the subject involved in the wide range of cognitive relations. Since, as Lycan points out, F is a logical fiction "if sentences which grammatically seem to refer to F's are (at the level of logical form) correctly analysed into sentences which refer, if at all, to entities of some different kind" (Lycan: 207), it seems clear that for Russell in the Empiricist phase the subject or self is *not* a logical fiction.

No doubt Lycan would reply to this by pointing out that while Russell does support dualism against neutral monism in Lecture IV of "The Philosophy of Logical Atomism" (PLA: 195–6), he admits that the arguments for the position no longer seem as compelling as they did in "On the Nature of Acquaintance" four years earlier. So much so that by Lecture VIII he appears to be more favourably disposed towards neutral monism (PLA: 240–2). I agree that this is so, nevertheless the closest that Russell comes to denying dualism is to admit that he doesn't "know whether neutral monism is true or not" (PLA: 242). Yet only according to neutral monism is the subject or self a "logical fiction" (PLA: 241), so Russell's stance is not enough to rescue (EL).

Lycan might also object that Russell talks as though he *is* dispensing with the subject:

> We shall not deny that there may be a metaphysical ego. We shall merely say that it is a question that does not concern us in any way, because it is a matter about which we know nothing and can know nothing, and therefore it obviously cannot be a thing that comes into science in any way. What we know is this string of experiences that makes up a person, and that is put together by means of certain empirically given relations such, e.g. as memory.
>
> (PLA: 240).

My response to this is that the science that Russell is here dealing with, which has no room for a metaphysical ego, appears to be *psychology*. This seems to follow from the way the example is set out by Russell. ("*What is it that makes you say* when you meet your friend Jones, 'Why, this is Jones'? . . . Jones is not constituted *as he is known* by a sort of pin-point ego that is underlying his appearances. . . . " (PLA: 239, my italics).) Taken in this light, a person is an epistemological fiction, but it doesn't follow that egos are logical fictions at the deeper levels of analysis underlying the specific sciences. After all, a few pages before the above passages, in a discussion of the analysis of the propositions of physics, Russell says that the aim is to find out "what is the smallest empirical apparatus — or the smallest apparatus, not necessarily wholly empirical — out of which you can build up these propositions" (PLA: 235). So why not non-empirical egos in an analysis of the general features of the world? Elsewhere the postulation of such non-empirical entities was preferred to the "logically impeccable" alternative of solipsism.[50]

Having maintained the falsity of (EL), the truth or otherwise of the second half of Lycan's first biconditional (i.e. (LE): All logical fictions are epistemological fictions, or contrapositively, all epistemological atoms are logical atoms) does not matter for present purposes. This is so because if (EL) is false, *a fortiori*, so is the first biconditional viz.: All and only logical atoms are epistemological atoms. That there is not such a close connection between the logical and the epistemological is indeed suggested by Russell himself near the start of "The Philosophy of Logical Atomism":

> The precise propositions that you arrive at may be *logically* premisses to the system that you build up upon the basis of them, but they are not premisses for the theory of knowledge. It is important to realise the difference between that from which your knowledge is, in fact, derived, and that from which, if you already had complete knowledge, you would deduce it. Those are quite different things. The sort of premiss that a logician will take for a science will not be the sort of thing which is first known or easiest known: it will be a proposition having great deductive power, great cogency, and exactitude, quite a different thing from the actual premiss that your knowledge started from.
>
> (PLA: 162).

These sentiments are repeated in the later passage that Lycan uses as an epigraph to his paper:

> . . . errors are liable to arise from assimilating the logical to the epistemological order, and also, conversely, from assimilating the epistemological to the logical order.
>
> (LA: 164).

Since the above two quotations belong respectively to the Empiricist and Modified Empiricist phases, they provide further support for the general point maintained in this chapter that primitive epistemological propositions don't coincide neatly with the premisses reached in logical analysis.

We have spent a lot of space on Lycan's first biconditional because he claims to have shown its truth. However he admits failure in his attempt to establish the second biconditional, viz. all and only epistemological atoms are ontological atoms. Accordingly we need not dwell on it, although it is probably worth pointing to a quick argument for its falsity. As Lycan defines them, ontological atoms are ultimate simples, i.e. lacking complexity. Epistemological atoms, again as Lycan defines them, have different degrees of complexity. So, the second biconditional is false. In any case, one worry is that Russell himself, wasn't as committed to ultimate simples as Lycan's definition assumes.[51] Another is that by setting up epistemological atoms as things known directly and immediately one covers a multiplicity of *types*. Even concentrating on sense-data as epistemological atoms there are difficulties, as Lycan notes, due to the complexity of sense-data.[52] However, things known directly and immediately also include facts[53] and universals,[54] so the second biconditional is not even remotely plausible.

Following the abortive attempt to bridge logical and ontological atoms via epistemological atoms, Lycan seeks to make the connection by what Russell says about a *logically perfect language*. One can agree with Lycan that in a logical language, as Russell describes it, logical complexity would mirror ontological complexity. However the point that needs to be stressed is that the logically perfect language, in which complexity of propositions mirrors complexity of facts, is an *unattained ideal*. The language of *Principia Mathematica* meets the *syntactical* requirements of a logically perfect language, but Russell was never able to meet the *semantical* requirements of such a language (PLA: 176).[55] In the lectures on the "Philosophy of Logical Atomism", Russell proposes to "*assume* that we have constructed a logically perfect language, and that we are going on State occasions to use it. . . . " (PLA: 176, my italics). This means, for example, that the subsequent discussion about 'this' and 'that' as logically proper names is a discussion of proper names in *English, not* in a logically perfect language. So while Russell held that a logical language in which the logical atoms were also ontological atoms was a possibility, it was never more than that. The *actual* progress of Russell's philosophical analysis in the Empiricist phase was somewhat less.

Arguing for the view that the premisses of Russellian philosophical analysis in the Empiricist phase were not simply, or even primarily, primitive epistemological propositions is the main thrust of this section. Having rejected some arguments to the contrary, we can more positively show a major case of Russellian analysis in the Empiricist phase agreeing with our account.

In lecture III of *Our Knowledge of the External World*, Russell sets out "to apply the logical-analytic method to one of the oldest problems of philosophy, namely, the problem of our knowledge of the external world" (*OKEW1*: 65). What follows reproduces the pattern made familiar to us in the Platonist phase. The first step is to identify the *data*, i.e. "matters of common knowledge, vague, complex, inexact . . . yet somehow commanding our assent as on the whole and in some interpretation pretty certainly true".[56] Russell's next step is to classify the data in such terms as primitive vs derived, logical vs psychological, hard vs soft (hard data are "those which resist the solvent influence of critical reflection" (*OKEW1*: 70). The upshot of this is to arrive at some unsurprising candidates for hard data: the facts of sense and the laws of logic, closely followed by facts of recent memory, etc. Even the inevitable acknowledgement of Descartes' influence is given. Just as important is the conclusion that certain common beliefs are not part of our hard data, viz. "that sensible objects in general persist when we are not perceiving them" and "the belief in other people's minds" (*OKEW1*: 72).

Having isolated the hard data, the problem of our knowledge of the external world becomes: "Can the existence of anything other than our own hard data be inferred from the existence of those data?" (*OKEW1*: 73). There follows a discussion for fourteen pages or so which attempts to provide an account of the world merely in terms of the hard data, i.e. taking what we have been calling "primitive epistemological propositions" as the premisses. When this turns out unsatisfactorily, Russell proposes to "make a new start, adopting a different method." He proposes, "in order to have a model hypothesis as a help for the imagination" to "construct one possible (not necessary) explanation of the facts" (*OKEW1*: 87). Russell proceeds then to outline his neo-Leibnizian three-dimensional world of perspectives, some of which are perceived by subjects. This constructed picture of the world fits the "facts of sense, the facts of physics, and the facts of physiology . . . there is no empirical evidence against it; it is also free from logical impossibilities" (*OKEW1*: 93). However, it also relies on testimony and the existence of other minds. Russell admits to being unconvinced by arguments for other minds, but nonetheless accepts the hypothesis that they exist on the grounds that "it systematises a vast body of facts and never leads to any consequences which there is reason to think false" (*OKEW1*: 96). In addition, ". . . it enables us to extend our knowledge of the sensible world by testimony. . . . " Russell's final conclusion regarding this analysis of our knowledge of the external world is that the

> . . . hypothetical construction . . . shows that the account of the world given by common sense and physical science can be interpreted in a way which is logically unobjectionable, and finds a place for all the data, both hard and soft. It is this hypothetical construction, with its reconciliation of psychology and physics, which is the chief outcome of our discussion. Probably the construction is only in part necessary as an initial assumption, and can be obtained from more slender materials by the logical methods of which we shall have an example in the definitions of points, instants, and particles; but I do not yet know to what lengths this diminution in our initial assumptions can be carried.
>
> (*OKEW1*: 97).

In the Empiricist phase, Russell never succeeded in reducing the initial assumptions behind this construction. That is to say, the premisses about other minds and sensibilia remained, thereby ensuring that in this phase, as in the Platonist one, primitive epistemological propositions were, at best, only part of the premisses reached by logical analysis.

Once again the contrast between the different interpretations is well illustrated by analysis diagrams. The view that primitive epistemological propositions constitute the premisses results in the following *foundationalist* diagram:

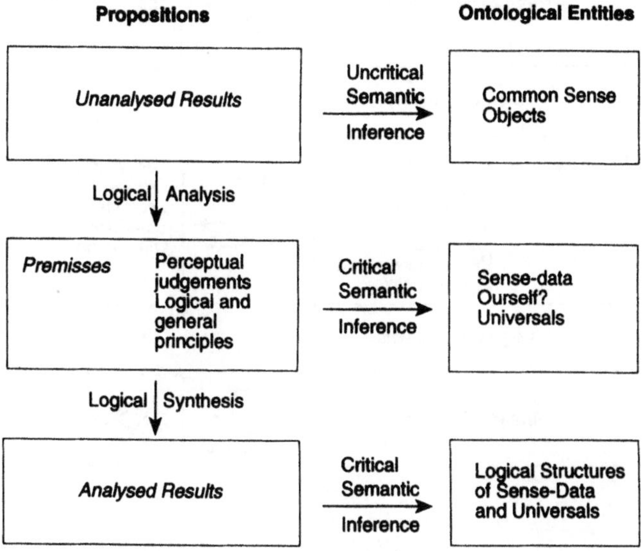

Figure 4.5.

Certainly this analysis diagram captures Russell's view that a

> ... complete application of the method which substitutes constructions for inferences would exhibit matter wholly in terms of sense-data, and even, we may add, of the sense-data of a single person, since the sense-data of others cannot be known without some element of inference.
>
> (RSDP: 12).

However, as with the logically perfect language, this attempt to "establish physics upon a solipsistic basis" was "an ideal, to be approached as nearly as possible, but to be reached, if at all, only after *a long preliminary labour of which as yet we can only see the very beginning*" (RSDP: 12, my italics).

The reality was that in the Empiricist phase Russell never got past the beginning, while the Modified Empiricist phase was initiated by the realisation that this solipsistic ideal would forever remain unattainable. Hence the actual state of Russellian analysis in the Empiricist phase is reflected in the following *non-foundationalist* diagram which shows the closest Russell ever came to a foundationalist position in his philosophical analysis. Yet the presence of propositions about inferred entities, such as sensibilia and other minds, in the premisses ensures that Russellian analysis in the Empiricist phase, as in the Platonist phase, complies with Table 3.1 in the previous chapter and the analysis diagram at the start of this chapter.

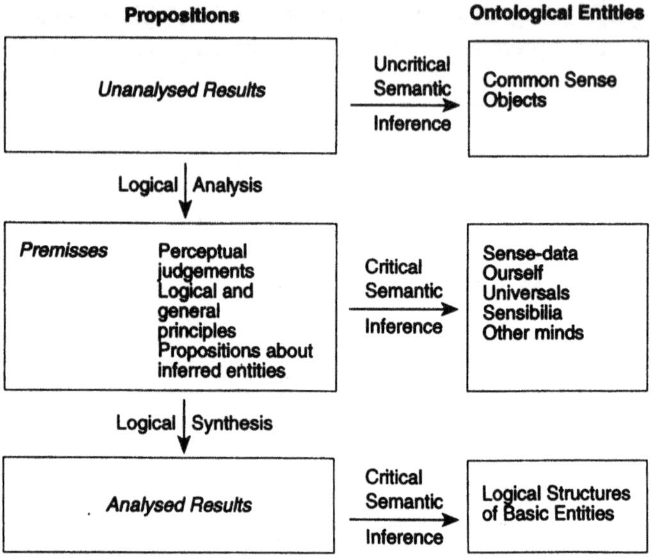

Figure 4.6.

4.5 Russell's Modified Empiricist Phase

It turns out to be a relatively simple matter to demonstrate that Russell's philosophical analysis in the Modified Empiricist phase, as in the two previous phases, complies closely with the features outlined in Table 3.1 in the previous chapter and the analysis diagram at the start of this chapter. This is so because the *essential characteristic* of the move to Modified Empiricism is the acceptance of the inescapable necessity of postulating inferred (or non-experienced) entities. By way of contrast, as we have seen, the Empiricist phase, though also marked by the postulation of such entities, involved the hope that such postulation could be avoided – a hope that we saw dashed. The inferred entities of the Modified Empiricist phase are *events*, i.e. "entities or structures occupying a region of space-time which is small in all four dimensions" (*AMa*: 286). The sense-data and sensibilia of the Empiricist phase become a mere sub-class of these events, with the added complication, that with the abandonment of the relational theory of sensation, the sensation just is the sense-datum, hence use of the latter term is largely abandoned.[57]

So, at the limit of analysis arrived at by Russell in the Modified Empiricist phase, events are the basic particulars which (together with universals) constitute the ontological entities assumed by the premisses (in the middle part of the analysis diagram). Of course, it is always possible that further analysis will lead to *other* basic particulars in respect of which events are no longer basic,[58] but for the moment it is events that are basic. It is obvious *prima facie*, that the analysis diagrams employing events comply with the relevant features of *results* and *premisses* detailed in Table 3.1 of the previous chapter. After all, if events are the only particulars assumed by the *premisses*, whereas the results (propositions of science, common sense, etc.) assume a multitude of kinds of particulars, then clearly the premisses have the potential to be simpler, more abstract, more precise, less obvious, etc., than the various results.

Before confirming these general claims by some actual examples, it is worth noting that the quasi-Cartesian sorting of propositions according to their degrees of self-evidence occurs again in the Modified Empiricist phase (RTC: 711–4), just as it did in the two preceding phases. As before, this procedure is a *preliminary* to analysis rather than the analysis itself. Once again this point is demonstrated by the contrast between *logical* and *epistemological* order. All of the foregoing elucidation of Russellian analysis shows it to deal expressly with *logical* order. That this is not the same as the *epistemological* order is clear from Russell's warning that:

> . . . errors are liable to arise from assimilating the logical to the epistemological order, and also, conversely, from assimilating the epistemological to the logical order.
>
> (LA: 164).

Not surprisingly, the results of *epistemological* ordering remain constant over the three phases. In contrast, the *logical* order changes, e.g. in the Platonist phase propositions about points are more fundamental than ones about sense-data; this order is reversed in the two later phases.

For a clear-cut confirmation that Russellian philosophical analysis in the Modified Empiricist phase is simply a continuation of what went before, we need look no further than the first chapter of *The Analysis of Matter*, a major book of the period. This work typifies Russell's concern with the foundations of physical science, a concern that lasted throughout the Modified Empiricist phase. In the first chapter of *The Analysis of Matter*, Russell outlines the *three* distinct stages involved in the philosophical analysis of physics. Stage one consists of a detailed consideration of physics, as the *results* to be philosophically analysed,[59] in which its various propositions are sorted into their *deductive* order. This procedure, which Russell calls the "logical analysis of physics" occupies all of Part I of *The Analysis of Matter*. Once again we are in danger of confusion about the term 'analysis'. Is this *logical* analysis of physics *philosophical* analysis or not? The answer is neither a definite 'yes' or 'no'. This is so because, as we found in our above discussion of the Platonist phase, today's premisses become tomorrow's results, so that as the frontiers of knowledge are pushed back, what was once philosophy becomes part of some other field of knowledge. Thus somewhere in prehistoric times it was a *philosophical* analysis to move from the result '2 sheep + 2 sheep = 4 sheep' to the premise '2 + 2 = 4'. However from our vantage point, such a move hardly belongs to philosophy, though it does involve a *logical* analysis of propositions. Certainly on Russell's view we have here *logical* analysis *within* mathematics, but no longer *philosophical* analysis as he understands it. The same goes for the analysis of physics in Part I of *The Analysis of Matter*. This logical ordering of the propositions of physics into a deductive system is seen by Russell as "a problem in pure mathematics" (*AMa*: 1) and "only a necessary preliminary to our main theme" (*AMa*: 6), which is, of course, the *philosophical* analysis of physics. So physics supplies the results to be philosophically analysed, as well as, perhaps, their logical ordering. But this dependence on others by the philosopher ceases when it comes to the *philosophical* analysis. As Russell put it, we:

> . . . need not listen to physicists outside physics, and it is for the philosopher rather than the physicist to ascertain just what physics asserts.
> (RTC: 701).

The outcome of this logical analysis of physics is that we reach "an initial apparatus of entities and propositions" (*AMa*: 1) from which the other entities and propositions can be defined and deduced respectively. These initial propositions are "hypotheses containing variables" and, in general, there will be "many different sets of initial hypotheses which will all yield the same body of propositions" (*AMa*: 2). The choice between the different sets of initial hypotheses is, of course, based on the usual criteria of economy and elegance. Philosophical interest begins with the various possible *interpretations* of the initial hypotheses, i.e. the substitution of some sets of objects for the undefined objects represented by the variables in the initial hypotheses. Usually, of course, there is one interpretation which is "much more important than the others" (*AMa*: 5), as Russell illustrates with his favourite example drawn from Peano's work. Peano's five axioms are verified by *any* progression, but the important interpretation involves the cardinal numbers (*AMa*: 4–5).[60]

The matter of *interpretation* brings us to the second stage in the philosophical analysis of physics, viz. "the application of physics to the empirical world". Since the evidence for the truth of the laws of physics is empirical, and since all empirical evidence is perceptual, "the world of physics must be, in some sense, continuous with the world of our perceptions" (*AMa*: 6). Thus, even though the two worlds are *prima facie* very different – indeed "physics and physiology themselves seem to give grounds for supposing that perception cannot give very accurate information as to the external world" (*AMa*: 6) – nonetheless the task is to find an interpretation of the initial hypotheses, arrived at in the logical analysis of physics, which "gives a due place to perceptions" (*AMa*: 7). This second stage in the philosophical analysis of physics is dubbed by Russell "the epistemological problem." If stage one was not philosophical analysis proper in the Russellian sense, then neither is the second stage on its own. Rather, stage two sets criteria which determine the ideal form that the analysis diagram should take and is thereby the *beginning* of the philosophical analysis. The form of the analysis diagram derived from stage two is as shown in Figure 4.7.

As can be seen from this analysis diagram, Russell's requirement for stage three will be that it should arrive at premisses that are common to both physics and perception, whilst at the same time involving no distinction between mental and material entities.

So Part II of *The Analysis of Matter* addresses "the epistemological problem"

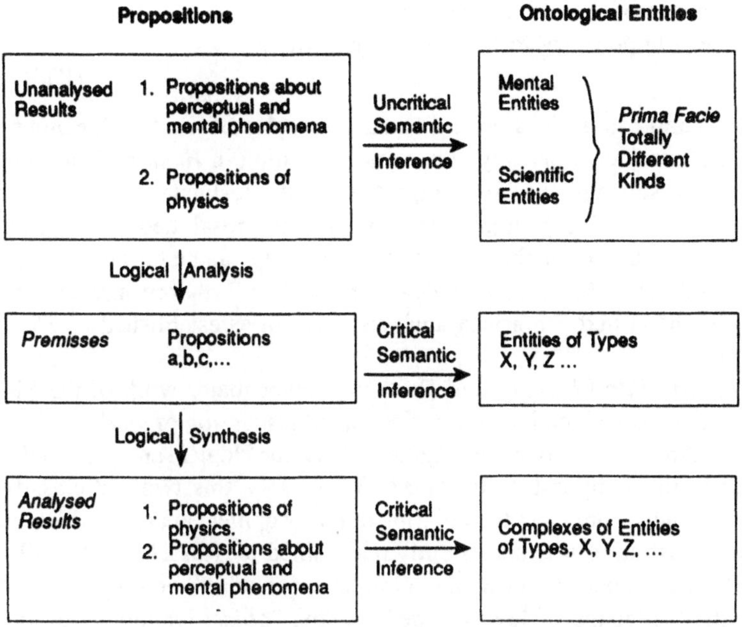

Figure 4.7.

and concludes that an analysis diagram of the above form will result if the premisses include the *causal theory of perception*, i.e., "the view that all our perceptions are causally related to antecedents which may not be perceptible" (*AMa*: 9).[61] The idea is that a knowledge of physics plus initial conditions would enable the explanation of and prediction of perceptual phenomena. This general causal account is intended to extend, with suitable modification, to other mental phenomena besides perception. Thus by Part III of *The Analysis of Matter*, Russell is set to complete the philosophical analysis of physics by seeking answers to the questions:

> What are the ultimate existents in terms of which physics is true (assuming that there are such)? And what is their general structure? And what are the relations of space-time, causality, and qualitative series respectively?
> (*AMa*: 9).[62]

The outcome of this third stage of the analysis is, as already noted, a set of premisses for the analysis diagram whose corresponding entities are restricted to *events* and *universals*. Clearly such an analysis diagram accords closely with the features of results and premisses summarised in Table 3.1 of the previous chapter.

Though the *ontological entities* corresponding to the *premisses* in *The Analysis of Matter* analysis diagram were never further analysed by Russell, he later became very interested in the detailed explication of premisses such as the causal theory, a task not attempted in *The Analysis of Matter*. So in 1943 he wrote that the

> . . . canons of scientific inference have never yet been formulated; if I have leisure, I hope to try to formulate them myself.
>
> (RTC: 718).[63]

Induction and causal lines were given as examples of what such canons would cover. This projected investigation came to fruition in *Human Knowledge*, Part VI of which sets out Russell's suggested set of "Postulates of Scientific Inference".[64] They are five in number and include the causal theory in a generalised form. Russell thought it "highly probably that they can be further reduced. . . . " (*HK*: 506).[65] Thus *Human Knowledge* represents a further refinement of details within the Modified Empiricist analysis diagram as established in *The Analysis of Matter.*

An Inquiry into Meaning and Truth is another major work of the Modified Empiricist phase which includes a significant discussion of *analysis*. In a chapter titled "Analysis", Russell distinguishes between "logical analysis" and "analysis into spatio-temporal parts" (*IMT*: 327). Does this twofold classification represent a departure from the doctrine of philosophical analysis which we have been demonstrating was consistently maintained by Russell? Not at all. In *Inquiry* Russell is concerned with the familiar notion of a "hierarchy of languages", where the "language of lowest type", i.e. the "*object-language*" or "*primary language*" has every word denoting "a sensible object or set of such objects. . . . " (*IMT*: 19). Languages of higher type are produced in the usual way by the addition of various logical and semantical words. Russell's interest in "analysis into spatio-temporal parts" arises in the context of an investigation of whether, in describing perceptual experience, the *primary language* can't avoid propositions of the form "P is part of W" (*IMT*: 336ff.). Russell's conclusion is that such propositions are essential in the primary language. "Logical analysis", on the other hand, "is that which proceeds from considering different kinds of words, and which inquires whether anything corresponds in the non-verbal world" (*IMT*: 330). Such analysis is best accomplished with the help of an artificial logical language and investigates "what there is" in the non-verbal world, not just in perceptual experience. In earlier parts of this chapter we have noted Russell frequently sorting propositions into *epistemological* order as a *preliminary* to philosophical analysis proper, which concerns *logical* order. We can now see that the "analysis into spatio-temporal parts" in the *Inquiry* belongs in the former category, i.e. the determination of what is *epistemologically* primitive. "Logical analysis" in the *Inquiry* however, belongs in the latter category, and, as Russell sees it as the more fundamental sort of analysis, it is now clear that the *Inquiry* is consistent with our overall account of Russellian analysis.

Throughout our discussions of the analysis diagram in this chapter, it has been evident that what the *premisses* are is relative to the stage that analysis has reached at any given time. Thus Peano's five axioms were *premisses* of mathematics when he postulated them, but relative to *Principia Mathematica* they are *results*. A useful way of talking about premisses, which is easily adapted to wider or narrower areas of knowledge, was introduced by Russell in the Modified Empiricist phase. The notion is that of *minimum vocabularies*:

> A minimum vocabulary for a given system of propositions is a set of terms having the two properties, (a) that no one of them can be defined in terms of the others, (b) that by means of all the terms, but not of any sub-class of them, all the propositions of the given system can be expressed.
>
> (RTC: 687).[66]

Thus, e.g., Peano reduced the minimum vocabulary of arithmetic to the *three terms* required by his five axioms. Frege, Russell and Whitehead held that the minimum vocabulary for mathematics is the same as for logic (MMD: 14–5). Clearly, at any given time, a reduction in the minimum vocabulary will lead to a revision of the premisses. But, says Russell:

> . . . broadly speaking, minimum vocabularies are more instructive when they show a certain kind of term to be indispensable than when they show the opposite.
>
> (MMD: 15).[67]

Such, for example, is the case with relation words. This is the major reason why universals are among the basic entities corresponding to the premisses in the analysis diagrams for each of the three phases. More will be said on this in the coming chapters dealing with relations. "In general, perhaps always," stresses Russell, "there will be many minimum vocabularies for any given system of propositions" (RTC: 687). As usual in such circumstances, simplicity and elegance are a useful guide. However, the possibility of a better minimum vocabulary being found is a major impetus for Russell's characteristic modesty about the finality of his analysis[68] – a striking contrast with the many *a priori* philosophers that preceded him.

We have seen enough to confirm that Russellian philosophical analysis in the Modified Empiricist phase has the same essential features that it had in the two earlier phases. Hence there is a single method of philosophical analysis exhibited by Russell throughout his post-idealist philosophy.

Conclusion

It is appropriate to conclude this rather lengthy chapter by emphasising two points, one directing attention to a significant trend that has emerged, the other connecting our findings about the three phases in Russell's realism to a major thesis of coming chapters. Firstly, we have seen that the search for *certainty* in both mathematics and science, and the failure of that search, have been major preoccupations underlying much of Russell's philosophising. The common view of this search for certainty had Russell trying to discover premisses that are certain, from which mathematics and science would be deducible. Failure, on this view, would consist in no such premisses being found. We have seen that this common view is *false*. Rather, as we found, Russell's search for certainty consisted in the attempt to arrive at sets of (far from certain) premisses from which the more certain propositions of mathematics and science would be deducible, thereby organising such knowledge and extending it. Failure to find the desired certainty in mathematics was largely due to Russell's discovery of the paradox about the class of all classes which are not members of themselves. This paradox was avoided by introducing restrictions into the premisses which led to ever increasing complications. So the problem was that a relatively simple, clear-cut set of premisses, from which the whole of mathematics, with *no* unwelcome paradoxes, could be deduced, proved to be elusive. Presumably Godel's 1931 incompleteness theorem, proving that all consistent axiomatic formulations of number theory include undecidable propositions, was a further blow to this enterprise.[69] In the case of science, the failure to find certainty appears to be due more to developments within science itself (i.e. the *results*), than to problems in arriving at suitable premisses. As Russell put it, "what modern physics has to say is somewhat confused. . . . " (*MPD*: 13), though he still held that it provided our best account of the world. This, then, is one aspect of Russell's loss of certainty.

Another is that certainty for Russell was closely tied to the notion of its being grounded in some extra-human reality.[70] So the quest for certainty in mathematics was prominent in the Platonist phase and the end of this phase heralded the end of mathematics as the focus of Russell's philosophising. Instead science as an account of supra-sensible reality, became the focus.[71] However, it wasn't long before Russell, influenced by physicists such as Eddington, was becoming increasingly oppressed by what he saw as solipsistic and subjectivist tendencies in modern physics.

> It seemed that what we had thought of as laws of nature were only linguistic conventions, and that physics was not really concerned with an external world.
>
> (*Auto.*: 393).[72]

While never quite accepting this view, it cast an ever present shadow over Russell's highest hopes for his endeavours. He met this alleged subjectivist trend in physics by arguing that the scope of scientific knowledge is much less than people have thought, being restricted to objective abstract "logical structures of events" so that "assuming physics true, what we know about the physical world is very little" (*HK*: 345).[73] This aspect of Russell's philosophy will be taken up in later chapters, but enough has been said, I think, to establish the claim that Russell's early optimism about justifying mathematics and science was somewhat dimmed by a series of unforeseen difficulties.[74] To be sure, as we have seen, Russell's notion of how to demonstrate certainty is somewhat *idiosyncratic*. Rather than the traditional view that has philosophy developing necessary general principles that are the basis of all thought, principles that confer on mathematics and science whatever certainty they possess, Russell takes a more modest stance. Contemptuous of the claims by philosophers to have the competence to assess the success or otherwise of mathematics and science (PTC: 461–2), Russell instead accepts that, at least in their broad outlines, they are the nearest thing we have to certainty. The role of philosophy is the subsidiary one of seeking general principles from which mathematics and science can be deduced, but with the difference that such principles are tentative and provisional. In addition, at any one time, there might be available more than one set of principles from which present mathematics or science can be deduced. The choice between the alternative sets of tentative principles depends, of course, on such matters as the success or otherwise of new propositions deducible from the principles. Thus Russell views philosophy as *inductive* in the same way that the various sciences are inductive (RMDP: 274–5 and 282).

In the two previous chapters three features of Russellian philosophical analysis were emphasised:

(i) Analysis is unlikely to be final;

(ii) Analysis enlarges the domains of particular subjects; and

(iii)Analysis leads to premisses that are decreasingly self-evident.

While (iii) has been dealt with extensively in the present chapter, (i) and (ii) clearly fit the account that has been given of Russell's preoccupation with certainty. (i) accords with the certainty residing more in the results than the premisses, while (ii) locates the current premisses at the frontiers of knowledge, i.e. in the realm of speculation rather than certainty. Russell's *territorial* metaphor for knowledge puts the certainty firmly in the centre well away from the frontiers, with the suggestion that as the frontiers expand, so also does the certainty they enclose.[75] In contrast the *bed-rock* metaphor goes naturally with the foundationalist view that he opposes, in which a *priori* certain premisses provide the philosophical underpinning of mathematical knowledge.

A major thesis to emerge from Part One of this book will be the claim that *Russellian analysis tends to proliferate relations*, i.e. relations tend to figure more and more in the analysed results as analysis proceeds. The precise reasons for this will emerge in the coming chapters on relations. In the meantime our findings about the analysis diagram for each of the three phases considered in this chapter can be seen as prefiguring this result. In each case the ontological entities corresponding to the basic premisses were *universals* (pre-eminently *relations*) and *particulars* (Platonic particulars, sense-data and events respectively). As we will find, relation words, more so than words for properties or particulars, seemed to Russell to be inescapably required in minimum vocabularies for mathematics, logic and science. Hence the complex entities corresponding to the analysed results in the analysis diagram will have relations as central components of the complex. This trend becomes more pronounced as analysis is pushed further.

Notes

1. See also PLA: 162–3 and LE: 134–6.
2. Strictly speaking, we cannot *know* that it exists and have no sound reason for assuming that it does exist.
3. In *IMT*, PU, *HK*, *MPD*, etc.
4. See also Quote E in Chapter Three.

5. See, e.g. LA: 173; *HK*: 268–9 and *MPD*: 164–5.
6. See e.g., Ayer (1971: 55ff.); Sainsbury (1979: 41–56); Brown (1978: Part 5); Lewis (1968: Ch. 5).
7. *IMP*: 142; PLA: 179–80; LA: 173; and *MPD*: 164, where Russell states that though he "on occasion spoke of atomic facts as the final residue in analysis . . . it was never an essential part of . . . (his) analytical philosophy . . . to suppose that such facts were attainable."
8. The "neat logical properties" here are such as being *eternal, independent, changeless*, etc., i.e. the properties traditionally attaching to Platonic entities such as numbers, points and instants. In contrast the properties of their replacement structures (classes of classes for numbers, complexes of sense-data or events for points and instants) are much less neat or familiar.
9. The Wood material is 25 pages or so of a projected detailed examination of Russell's philosophy cut short by the author's death. Russell thought that it "admirably clarifies" difficult aspects of his work and included it at the end of *My Philosophical Development*, adding in a prefatory note that, had it become available sooner, he would have put it at the beginning.
10. On the failure to find certainty in mathematics, see *Auto.*: 395. On the same failure in science, see *HK*: 527.
11. See e.g. Hamlyn, 1970: 10–11.
12. Besides Lakatos, see C. Fritz Jnr, "Russell's Philosophy of Science" in Pears (ed.) (1972), where he claims (p. 148) that "Russell is looking for knowledge that is 'basic' in the sense of being 'certain' (or almost certain) which can serve as premises for the remainder of our beliefs." See also, e.g. Wood, 1957: 48 and Lewis, 1968: 19.
13. The following quotations are from Lakatos, 1978b. See also Lakatos, 1976: 56–7, especially the editors' note.
14. It might be objected that this fourfold division overlooks significant developments *within* phases, e.g. the 1905 theory of descriptions is often seen as a new departure. Answers to this sort of objection are given later in this chapter. Also Part Two of this book provides a detailed rationale for the four phases.
15. For more on the 'Tiergarten Programme' see Griffin, 1991: 79–88.
16. For details see Griffin, 1991: 35–42.
17. For details see Griffin, 1991: Chapter 8.
18. For details see Griffin, 1991: 81–4.
19. What is missing, of course, in the neo-Hegelian phase is the later synthesis step in which analysed results are constructed from the premisses. In fact, there were several forerunners to what became the synthesis step in Russellian analysis. Before 1898, he employed the second stage of a two-part transcendental deduction. Then from 1898–1900 Russell sought to develop a new approach based on part-whole relations. (See, e.g., FIAM: 291–5.) After 1900, under the influence of Peano's work, the full-blown method of analysis was developed.
20. Moore (1988) argues that Russell's neo-Hegelianism provided valuable training in paradox hunting that was to bear fruit in his later work. He also emphasises the important influence of Cantor's work on Russell's thought about paradoxes.
21. Moore's influence on Russell encouraged him to take the desubjectification of Kant even further (see Griffin, 1991: 304–6). The desubjectification of Kant was also a concern of the German neo-Kantians in the mid-19th century.
22. As will be demonstrated in Chapter Ten.
23. In a letter to Gilbert Murray, 12 July 1911, Russell wrote: "There is a fair amount of metaphysics, but there seems to me so little that is ascertainable in metaphysics that I felt I must make theory of knowledge the framework. However, the two mix: Platonic ideas, e.g., which I believe in and expound, are very much on the border-line" (quoted in *Papers* 6: xli–xlii). Likewise in "Analytic Realism" (1911), in which Russell was developing the ideas that became logical atomism, he still talked in Platonist terms, e.g. "The atoms are of two kinds: universals, which are similar to Platonic ideas, and particulars. . . . " (AR: 133–4) and "the exactitude of mathematics depends on its being concerned with an abstract world, a world of

essences or Platonic ideas, which is not subject to the inaccuracies of sensory knowledge . . . . " (AR: 134).

24. See Griffin (1993a: 240ff.) for details.
25. "We have acquaintance with sense-data, with many universals, and possibly with ourselves, but not with physical objects or other minds" (KAKD: 161). Since "every proposition which we can understand must be composed wholly of constituents with which we are acquainted" (KAKD: 154), it follows that, in the final analysis, propositions apparently about other things, must be about sense-data and certain universal concepts.
26. This broad summary is of Russell's public position and omits details contained in the ferment of papers that he produced in this period as he tried various ways to render his theory consistent. (Many of these papers remained unpublished until they appeared in *Papers*.) For discussion see, e.g., Griffin, 1993a; Landini, 1987 and 1991; Rodriguez-Consuegra, 1989a, 1989b and 1991.
27. By 1901, there is a denoting concept for each number which appears to denote it and is easily confused with it.
28. See, e.g., Griffin (1985) for discussion of this.
29. Cf. ". . . logic is concerned with the real world just as truly as zoology, though with its more abstract and general features" (*IMP*: 169). For some possible difficulties with the ontology of Russell's theory of terms see Grattan-Guinness, 1976: 162–3, and 1980: 74–5.
30. Russell's various attempts to deal with his paradox were, of course, a major preoccupation throughout the Platonist phase. Prominent landmarks include the *Principles of Mathematics* theory of types, several years working on methods that didn't necessarily assume a type structure, the theory of descriptions, the 1906 'no-class' theory, and then the construction of the ramified theory of types of *Principia Mathematica*.
31. Russell's theory of terms, on which the theory of denoting in *Principles of Mathematics* is based, had a precursor in Moore's theory of concepts (Moore, 1899). The Russellian theory of terms, as non-linguistic items, appears in his writings for five years prior to publication of *Principles of Mathematics*, particularly in AMR.
32. Since 'term' was intended to be the widest word in Russell's philosophical vocabulary, denoting concepts are, of course, themselves terms. So the oft-quoted passage from *Principles of Mathematics* is consistent with both interpretations. Griffin further points out that the letter to Victoria Welby shows *Principles of Mathematics* to have a four-tier semantic theory of words, ideas, concepts and denotations as against the three-tier semantic theory of words, concepts and denotations that underpins the common interpretation.
33. Cocchiarella adds that the fact that "the new practitioners of analytic philosophy have in general failed to recognise or appreciate this similarity . . . " is in part ". . . the result of an inadequate understanding of Russell's early philosophy. . . . "
34. See also SOM: 91–2 for more on the "non-human" character of mathematics.
35. The general point is discussed by Lakatos on pp. 10–20 and 25ff.
36. Misquoted by Lakatos, 1978b: 25.
37. *PoM*: 106 gives the full list of indefinables.
38. For all I know, Lakatos' response to all of this might well have been that he was *not* writing real history, but rather a *rational reconstruction* of history (as in I. Lakatos (1978a): 53, 55, 61 and 120). However we are interested in Russell's *actual* views, not what they *ought* to have been according to Lakatos' theory of the history of mathematics.
39. RMDP was read before the Cambridge Mathematical Club, 9 March 1907 and first published in *EA*, 1973. Lackey's comment is from his editorial introduction to Section V of *EA*: 254.
40. E.g. Russell illustrates the general principles by considering Peano's and Frege's analyses of mathematics, much as was done in Chapter Two above.
41. See also *PP*: 70ff.
42. This passage was cited by Demopoulos (1981: 166) in support of his criticism of Griffin's (1980) account of Russell on the foundations of mathematics. In response, Griffin (1983: 43) admitted that the passage refutes his earlier suggestion "that Russell sought to demonstrate the certainty of mathematics by deduction from axioms which are certain." However he also

points out that the passage leaves untouched his "wider claim that Russell was more concerned in *Principia* and later writings with the certainty of mathematics rather than its generality." Griffin's considered conclusion that "Russell was attempting to establish the certainty of mathematics through a transcendental deduction of its axioms" is very close to our account of Russell's position in his Platonist phase. For discussion of the generality of mathematics in Russell's logicism see Grattan-Guinness (1990).

43. See also pp. 112 and 117.
44. E.g. see SMP: 69; RSDP: 12; *OKEW*, Preface; *PoM*, 1937 Introduction.
45. E.g. see Brown, 1978: Part 5.
46. Lycan understands the various sorts of atoms as follows: *Logical atoms* are the referents of linguistic expressions which genuinely refer at the limit of Russellian analysis, i.e. logically proper names. *Epistemological atoms* are things which are known directly and immediately, i.e. not by inference from prior knowledge of entities of any other kind. *Ontological atoms* are the ultimate simples or basic building blocks having no proper constituents.
47. Both passages are quoted in Lycan, 1981: 209.
48. E.g. *PP*: 50 and KAKD: 149.
49. E.g. see PLA, Lecture IV and ONA, Section II.
50. *PP*: 22ff. and RSDP: 12. For later reactions to solipsism see *HK*, Pt. III, Ch. II and *Auto.*: 393.
51. PLA: 179–80, LA: 173 and *MPD*: 164. See also above footnote 7 (this chapter) and Griffin, 1982: 74. Lycan (footnote 24 of his paper) notes Russell's PLA and LA doubts about ultimate simples, but views them as passing doubts.
52. KAKD: 149; *PP*: 101–2; NSD: 184; *OKEW1*: 114.
53. PLA: 175–7; ONA: 127. This also applies in the Platonist phase (see KAKD: 149–50 and NSD: 184).
54. ONA: 127. This also applies in the Platonist phase (see KAKD: 149–50).
55. See also Russell's 1922 Introduction to Wittgenstein's *Tractatus Logico-Philosophicus* where the logically perfect language remains an unattained, and probably unattainable, ideal.
56. The full quote is D in the previous chapter.
57. See, e.g. *MPD*: 100ff. LE is a case of Russell still using the term 'sense-data', though not because of any shift of doctrine.
58. See *MPD*: 15 and *AMa*: 286.
59. Cf. RTC: 700: "In ontology I start by accepting the truth of physics."
60. See also *IMP*: 8ff. and 24ff.
61. See also *MPD*: 17ff.
62. "Qualitative series" are those formed by colours, musical notes, etc.
63. Cf. MMD: 16.
64. See also *MPD*: Ch. 16.
65. See also *MPD*: 149.
66. See also MMD: 14–6 and *HK*, especially Part IV.
67. See also RTC: 687–8.
68. E.g. *AMa*: 10; RTC: 690; *MPD*: 151.
69. See *Auto.*: 672, "The followers of Gödel had almost persuaded me that the twenty man-years spent on the *Principia* had been wasted and that the book had better been forgotten."
70. SOM: passim; *MPD*: 118–9; MMD: 12–3.
71. In the Empiricist phase, unsensed sensibilia and other minds constitute supra-sensible reality.
72. See also: 395 and *AMa*: 6.
73. See also *MPD*: 13; *AMa*: 10.
74. On this point see A. Wood in *MPD*: 194–5.
75. Russell's metaphor is similar to Popper's *swamp* one in which piles are driven deep enough to carry the structure without ever reaching bed-rock. (See Popper, 1968: 111.)

CHAPTER 5

Relations in Mathematical Philosophy

5.1 The Indispensability of Relations for Russell

According to Russell, "the importance of relations" is "a matter which concerns philosophy and mathematics in equal measure".[1] His book on Leibniz "had emphasised the importance of relational facts and propositions as opposed to facts consisting of substance-and-attribute and propositions consisting of subject-and-predicate." Russell found that "the prejudice against relations had had bad consequences in mathematics as well as in philosophy." In mathematical logic the tendency was to emphasise classes, with relations at best being treated as classes of couples. For Russell it was a disaster for mathematical logic that technical considerations had been allowed to obscure the philosophical importance of relations. "What is important in the logic of relations is what is different from the logic of classes. . . . " Just what the important philosophical principles are that derive from the logic of relations will unfold as this chapter proceeds.

Russell himself had become gradually aware of the importance of relations for mathematical philosophy in the works that ended his idealist phase. As discussed earlier, Russell's neo-Hegelian project was to analyse the nature of the various sciences and the dialectical relationships between them. By the time that he abandoned idealism, he had completed detailed work on geometry, physics and pure mathematics. In each case, various paradoxes or antinomies were identified. During this work Russell adhered to a doctrine of internal relations, which held that all relations are internal, and also that each different relation is grounded in distinct intrinsic properties.[2] As already noted in Section 4.2.2, the paradoxes that Russell found while working on his dialectic of the sciences are essentially of two different kinds, which Griffin calls 'paradoxes of relativity' and 'paradoxes of parts and wholes'. It turned out that paradoxes of relativity assume the doctrine of internal relations, while paradoxes of parts and wholes arise from a false hypostatization of relations. Hence, "the doctrine of internal relations is responsible for both kinds of antinomy in Russell's dialectic of the sciences" (Griffin, 1991: 323).

After abandoning idealism, Russell always pointed to the doctrine of internal relations as the central plank of neo-Hegelianism. However, as Griffin (1991) shows, it took much patient and detailed work by Russell before it became apparent that the doctrine of internal relations was indeed the common root of the various paradoxes that he had described. In the meantime, the irreducibility of asymmetrical relations, a thesis that directly contradicts the doctrine of internal relations, gradually came to Russell's attention as he attempted to develop a neo-Hegelian account of signed arithmetic. In 1898 he worked extensively on the concept of sign in mathematics.[3] This work resulted in a rash of antinomies of sign and Russell came to see that both the concept of sign and the antinomies of sign depend upon relations. This in turn served to direct Russell's attention to the notions of order and of asymmetrical relations. The result of the foregoing was that by 1899, he finally came to reject the doctrine of internal relations. In doing so, of course, he adopted the doctrine that the ordering relations between (say) points are external, which means that the relations between the points are real and not reducible to properties of the points, nor are the points themselves altered in any way by being so related. Russell's neo-Hegelian views on relations will be discussed further in the next chapter.

Following the abandonment of idealism, Russell's work in mathematical philosophy centres on the project of logicism. This is the thesis that mathematics derives from logic in that its principles can be deduced from principles of logic and that its concepts can be defined in purely logical terms.[4] Philosophical analysis is applied to mathematics (the *results*) in order to arrive at logic (the *premisses*) so that at the successful completion of analysis, mathematics is deducible from logic. Russell's first attempt at this programme of analysis was *Principles of Mathematics*, followed later, and more satisfactorily, by *Principia Mathematica* (in collaboration with A.N. Whitehead). In the course of this analysis, Russell found that a "great part of the philosophy of mathematics is concerned with *relations*" (*IMP*: 42), i.e. relations often appeared in the premisses from which mathematics was deduced. Thus analysis reveals that the "most elementary properties of numbers are concerned with one-one relations, and similarity between classes" (*IMP*: 195–6). Multiplication turns out to involve a certain kind of one-many relation called "selections". Finitude involves ancestral relations, and so on for large areas of mathematics.

The importance of relations in the premisses of mathematical philosophy is evident from Russell's statement of the central thesis of *Principia Mathematica:* "that from certain ideas and axioms of formal logic, *by the help of the logic of relations*, all pure mathematics can be deduced, without any new undefined idea or unproved propositions" (LA: 163, my italics). (Cf. Whitehead's 1917 statement that, assuming logic, "mathematics is developed from the concepts of class and correlations" (1970: 170), where "correlation" is his general term for the various sorts of relations.)

Of course, at this stage, we are faced by the standard objection that relations aren't really of fundamental importance for mathematical philosophy because they can be reduced to classes of a certain sort. Naturally, Russell was well aware of this objection, yet was unmoved by it. As we shall see, his reasons for this stance are crucial to the overall shape of his philosophy. Obviously, Russell wasn't opposed to the *principle* of analysing relations in terms of something else. Indeed, he is committed to today's premisses becoming tomorrow's results as the frontiers of knowledge expand. Yet although he discovered or accepted analyses of *particular* relations,[5] Russell never found a satisfactory general way of analysing relations. This is one reason why, notwithstanding their neglect by most philosophers, relations are an essential aspect of all of Russell's philosophy.

As noted, Russell was appalled that the technical possibility of reducing relations to ordered pairs had obscured their importance. He had a variety of arguments against the view that formal reduction settled the matter — arguments that he still thought convincing late in his career (*MPD*: Ch. 8) when mathematical logic had developed even further from a technical point of view, e.g. the Kuratowski device and the Wiener device, each of which allows a formal reduction of ordered pairs to classes.[6] For a start, when a relation is reduced to a set of ordered pairs, according to Russell it is the *intension* alone which gives unity to the set. We understand the relation concept, e.g. 'precede', even though we cannot enumerate all of the pairs of terms related by the relation. That is, we are acquainted with the relation 'precede' without being acquainted with all of the elements of all of the classes which form the relation on the Kuratowski and Wiener definitions.

Likewise the Kuratowski and Wiener devices don't dispense with the need for intensions. The point of this argument by Russell is that while it is true that in a *formal* language we can *choose* to represent relations in the manner of Kuratowski or Wiener, in his view such a language would be defective in that it would fail to make essential distinctions. After all, on a standard interpretation of class existence, the class of classes {{a}, {a, b}}, (say), exists whenever a and b exist, *whether* a *and* b *are related by R or not*. According to Russell it makes, in many cases, an *empirical* difference whether a and b are related by R or not, a difference not made in the formal language. The user of the formal language has to rely on the intension of the relation to pick out the required classes, and so has not dispensed with relations.

Russell's next argument attacks the first step of the alleged reductions of relations. The fundamental claim of the argument is that ordered pairs can't serve as a reduction of relations because *order* presupposes relations. The

> . . . couples have to be *ordered* couples, that is to say, we must be able to distinguish the couple *x, y* from the couple *y, x*. This cannot be done except by means of some relation in intension. So long as we confine ourselves to classes and predicates, it remains impossible to interpret order or to distinguish an ordered couple from a class of two terms without order.
>
> (*MPD*: 67).

For Russell order is fundamental to both mathematical philosophy and general philosophy and no satisfactory method is available for reducing the asymmetrical, transitive, connected relations which generate order.[7]

Because of Quine's influence in leading the "flight from intension", Russell's first two arguments, however persuasive, would not convince a lot of philosophers. However, the argument from order is not so easily ignored, a view supported in a very suggestive paper by Vuillemin (1972). According to Vuillemin, Russell's theory of order represents a solution to "the problem of ontological commitment", a solution that provides a viable challenge to Quine's famous response to the same problem. Vuillemin begins by considering Russell's early development of a "platonist *principle of abstraction*" in reaction to Peano's "nominalist *definition by abstraction*".[8] The latter is nominalist because it involves propositions dealing with individuals only, whereas the former, devised by Russell in response to the defects in Peano's work, is platonist in that it involves the existence of universals. However, this early response by Russell to Peano was quickly revised when it became apparent that the principle of abstraction generates some of the class paradoxes.

Hence as Vuillemin shows, the theory of types led Russell to adopt a more nominalist interpretation of the principle. Thus by 1914 Russell was speaking of

> . . . 'the principle of abstraction' . . . which might equally well be called 'the principle which dispenses with abstraction' and . . . clears away incredible accumulations of metaphysical lumber. . . .
>
> (*OKEW1*: 42).[9]

In his platonist interpretation the principle led Russell to infer the existence of a common *property* in any case of symmetrical and transitive relations. However the nominalist interpretation holds that when

> . . . a group of objects have that kind of similarity which we are inclined to attribute to possession of a common quality . . . membership of the group will serve all of the purposes of the supposed common quality . . . therefore, unless some common quality is actually known, the group or class of similar objects may be used to replace the common quality, which need not be assumed to exist.
>
> (*OKEW1*: 42).

The nominalist turn here is distinctive – the principle of abstraction became just one more example of Occam's Razor in Russell's philosophical kit-bag (LA: 164).

It might appear, then, that all overtones of platonism have been banished from the principle of abstraction. Not so, according to Vuillemin, because the principle is integral to Russell's logic of relations and platonism about relations survives the demise of platonism about properties. In the logic of relations,

> . . . the principle of abstraction consists in replacing an equivalence relation, and a partial ordering on a set of elements, by an asymmetrical one-one relation between these elements and their equivalence classes, which themselves become totally ordered.
>
> (Vuillemin, 1972: 315).

Hence, we are back to the asymmetrical, transitive, connected relations which generate order and which seemed to Russell to have ontological import. So Russell's nominalist turn is still a battle within *intension*.

According to Vuillemin, against Quine's "to be is to be the value of a variable" criterion of ontological commitment, Russell holds that "to exist abstractly is to impose an order and a sense on elements, which do not contain either in themselves (theory of asymmetrical relations)" (Vuillemin, 1972: 320).[10] Vuillemin then compares the strengths and limitations of the two accounts and concludes that there is no clear decision either way. Quine's criterion is in line with certain requirements of Gödel's criterion of completeness, whilst Russell's isn't. However, Russell's criterion complies with Church's criterion of decidability, whilst Quine's doesn't.

Whatever the force of these metalogical points, the major merit of Vuillemin's paper from our viewpoint is that it clearly compares the Russellian and Quinean ontologies, a comparison that reveals that Russell lined up against Quine doesn't look nearly as unsophisticated as many philosophers might have expected. Quine's ontology consists of particulars and classes with relations and properties being reducible, whilst Russell is left with related particulars as fundamental, properties and classes being reducible. For Russell particulars are reducible to simpler related particulars, perhaps endlessly so. Thus reduction of related particulars only leads to more of the same kind. Russell's position has been generally ignored because relations are metaphysically problematic, yet how is the Quinean alternative an improvement? Apart from the well known difficulties with classes,[11] the reduction of relations to ordered pairs is very unsatisfactory. As Vuillemin (1972: 319–20) sees it, Quine's version of Wiener-Kuratowski type reductions involves quantification over universals, and offends against Church's theorem by obliterating all distinctions between classes and relations.

I conclude that Russell's relation ontology at least deserves a better hearing. I will argue that platonism regarding relations is much more significant in Russell's philosophy than has been hitherto acknowledged.

5.2 Special Characteristics of Relations

To elucidate more fully how order and relations in general were foundations of Russell's wider philosophy, we need to consider the ways in which he claimed that the logic of relations differed from the logic of classes. Summarising Part 1, Section D, of *Principia Mathematica*, Russell and Whitehead state that it

> . . . deals with those properties of relations which have no analogues for classes. . . . Most of the properties of relations which have analogues in the theory of classes are comparatively unimportant, while those that have no such analogues are of the very greatest utility. It is partly for this reason that emphasis on the calculus-aspect of symbolic logic has proved a hindrance, hitherto, to the proper development of the theory of relations.
>
> (*PM*, Vol. 1: 88).

What then are these unique properties of relations which Russell and Whitehead found to be "constantly needed throughout the rest of the work"? Three sorts of relations are said to be "specially important" (*PM*, Vol. 1: 299–301).[12] These are:

(i) Relations that give rise to descriptive functions;
(ii) Relations that establish correlations between different classes; and
(iii) Relations that generate series.

We will consider each of these in turn.

Relations which only one term at most can have to a given term generate descriptive functions. Such *one-many* relations are common amongst mathematical functions, e.g., 'the sine of *x*', 'the logarithm of *x*', etc. In general, a descriptive function is 'the term having the relation *R* to *x*', or, more simply, 'the *R* of *x*', where *R* is any one-many relation. Some of the most central concepts in the logic of relations are themselves descriptive functions, e.g. converse, domain, converse domain, field. Frequently relations which don't, as they stand, give rise to a descriptive function, do so when their domain or converse domain is suitably limited. The relation 'parent' is a case in point – it becomes a one-many relation that generates descriptive functions when its domain is limited to males or limited to females.

The second sort of "specially important" relation establishes a correlation between two classes. Such relations are one-one and occur in cases where at most one *x* has the relation *R* to a given *y and* also there is at most one *y* to which a given *x* has the relation *R*. (This is really a particular case of relations giving rise to descriptive functions, viz. the case in which the converse relation also gives rise to a descriptive function.) Relations establishing correlations are of fundamental importance to arithmetic. They are used in defining what is meant by saying that two classes (or series) have the same cardinal (or ordinal) number of terms. A more mundane example of such relations is the relation of marriage in a society where there is no polygamy.

Finally, in this trinity of important relations, are those that give rise to *series*. A series is a set of terms having an *order* derived from a relation having the three properties of being asymmetrical, transitive and connected. (A relation is *asymmetrical* when if *x* has the relation to *y*, then *y* does not have it to *x*; it is *transitive* when if *x* has the relation to *y* and *y* has it to *z*, then *x* has it to *z*; it is *connected* when if *x* and *y* are any two different terms in its field, then either *x* has the relation to *y* or *y* has it to *x*.) Any relation having these three properties arranges the terms of its field in a series. The prevalence of series in mathematics shows the importance of such relations to mathematical philosophy. Also common are relations that have "more or less of a serial character, without having all the characteristics necessary for generating series" (*PM*, Vol. 1: 300).[13] In such cases it is possible to devise cognate relations which are serial in the full technical sense. Thus, for example, the relation of son to father can be used to generate a multitude of series. Starting with me it can generate a series which ends at Adam; starting with my brother it can likewise generate a *different* series ending at Adam. Yet neither series orders the *field* of the relation of son to father, since both my brother and I belong to the field of the relation and one of us is missing from each of the series. In any case, the relation of son to father is not serial in the full technical sense because:

a) it is not *transitive*, i.e. I have the relation in question to my father, as he in turn has the same relation to my grandfather, but I *don't* have the relation to my grandfather; and
b) it is not *connected*, i.e. both my brother and I are terms in the field of the relation, yet neither of us has the relation in question to one another.

However, using techniques developed in *Principia Mathematica*, a new relation can be defined which is serial in the full technical sense and has for its field *x* and all of his ancestors in the direct male line (*PM*, Vol. 1: 301).[14]

These, then, are the three sorts of relations said by Russell to be "specially important" for philosophy. It is noteworthy that not only are the three sorts of relation *not* mutually exclusive, but it is possible for the one relation to have *all* three characteristics (*PM*, Vol. 1: 301).[15]

Do these three sorts of special relation give us any clues to what Russell and Whitehead had in mind when they stated that the logic of relations is important where it differs from the logic of classes? Yes, they do. Of vital importance in Russell's general philosophy is a distinctive feature of serial relations which is apt to be overlooked especially by those of a nominalist turn of mind. Whereas the *class a, b, c* is constituted by its members and nothing more, the *series a, b, c* is constituted by its members *plus* an ordering relation. To put it another way, the ontology of a class is exhausted by its members, but the ontology of a serial relation is not exhausted by its field. Russell argues for this point (*IMP*: 34) by considering the following six different series:

1,2,3; 1,3,2; 2,3,1; 2,1,3; 3,1,2; 3,2,1.

Clearly, the field of the serial relation is the same in each case, yet we have six *different* serial relations, so the relation is something other than the field. "Given the ordering relation", concludes Russell, "the field and the order are both determinate. Thus the ordering relation may *be* taken to the series, but the field cannot be so taken" (*IMP*: 34).[16]

This difference between classes and serial relations can also be put in terms of the 'logical fiction' concept much favoured by Russell from 1914 onwards. Classes *are* logical fictions, because they are not some further thing additional to their members, whereas serial relations are obviously *not* logical fictions. This point was invoked by Russell when considering the *formal* possibility of reducing all other relations to one-many relations. The asymmetrical relation *less than*, for example, can be formally replaced by *being a member of the proper ancestry of.* Thus *less than* is reduced to a one-many relation in which the one is a class, together with membership of that class. Concerning this method of reduction, Russell's conclusion was that ". . . there is every reason to think that it does not represent a philosophical analysis, if only because classes must be regarded as 'logical fictions'" (*IMP*: 45). Clearly Russell had in mind here the vital differences between the logic of relations and the logic of classes.

> Asymmetrical relations are . . . the most characteristically relational of relations, and the most important to the philosopher who wishes to study the ultimate logical nature of relations.
>
> (*IMP*: 44–5).

Since such relations are the basis of serial relations, which are *not* logical fictions, they cannot be reduced to classes which *are* logical fictions.

As will be demonstrated in detail in this book, the ontological import of serial relations is a doctrine that Russell maintained throughout his philosophical career from the onset of his revolt against idealism. It is a doctrine that is vital to a proper interpretation of his philosophy, yet it is also a doctrine that has been widely misunderstood or ignored.

5.3 Confusions About Relations

As happens with depressing frequency in Russell's works, he himself contributes to the misunderstanding by writing in a way that is bound to confuse the casual reader. Thus, only a few months before writing about the distinctive difference between classes and serial relations discussed above, Russell asserted in his "Philosophy of Logical Atomism" lectures that "series and classes are of the nature of logical fictions" (PLA: 171).[17] Here surely is a case of Russell's notorious chopping and changing his philosophical views? Between early 1918 and mid-1918 he obviously ceased to think that serial relations are logical fictions. Such would be a common reaction, yet, as frequently happens in such cases, closer inspection reveals that there is no conflict of views at all.

Thus, Russell's statement that "series and classes are of the nature of logical fictions" is combined in his "Philosophy of Logical Atomism" lectures with a nominalist view of classes (228ff.) and a realist view of relations (234). The apparent conflict of views is easily explained by Russell's desire to disown the exuberant realism of his *Principles of Mathematics* period. In *Principles of Mathematics* he developed a *class as one* and *class as many* doctrine. The *class as one* is "a whole of a particular kind . . . an *aggregate*" (*PoM*: 139). It is a particular *additional to* the particulars which make up the class, i.e. the *class as many*. A series, on the other hand, involves *two* kinds of wholes. As with classes, the *aggregate* of its members constitutes a whole, but in addition "there is another kind of whole, which may be called a *unity*" (*PoM*: 139). Whereas an ". . . *aggregate* . . . is completely specified when all its simple constituents are specified; its parts have no direct connection *inter se*, but only the indirect connection involved in being parts of one and the same whole" (*PoM*: 140), this is not so for a *unity*. Unities

> . . . contain relations or what may be called predicates, not occurring simply as terms in a collection, but as relating or qualifying. Such wholes . . . are not completely specified when their parts are all known.
>
> (*PoM*: 140).

Because Russell had a simplistic theory of meaning in *Principles of Mathematics*, whereby each word must have meaning in isolation, he is able to use the concept of a proposition to distinguish between aggregates and unities. A ". . . *unity* . . . is always a proposition, though it need not be an *asserted* proposition" (*PoM*: 139),[18] whereas an *aggregate* is a mere collection of terms. Russell's example is the case of *A* being greater than *B* – the parts *A, B* and *greater than* "compose simply an aggregate, or either of the propositions '*A* is greater than *B*', '*B* is greater than *A*'". This led Russell in 1903 to worry about "analysis as falsification" since it reduces propositions to aggregates, i.e. while the parts of the aggregate are also the parts of the unity, the *way* the parts were put together is lost when we analyse the unity into the corresponding aggregate. Russell's conclusion then (in 1903) was that "though analysis gives us the truth, and nothing but the truth, yet it can never give us the whole truth" (*PoM*: 141). Though some of these doctrines were to be later abandoned, due to such developments as a more sophisticated theory of meaning and the theory of types, the worry about the analysis of relational propositions was, as we shall see, to persist.

The point of the above discussion is to clarify alleged inconsistencies in how Russell viewed serial relations. As we have seen, in 1903 the ontology of a class comprised the individual members (the class as many) plus a further individual, viz. the aggregate of its members (the class as one). The ontology of a series comprised three components, viz. the individual members, the aggregate of the members, and the unity of the members. As was discussed in Chapter Four, Russell later moved to a more economical view: the ontology of a class became its members only, the class and aggregate being *logical fictions*. In the case of series, the ontology was reduced to the members of the series *plus* the ordering relation or unity, the series as aggregate being a logical fiction.

So when Russell states that "series and classes are of the nature of logical fictions" or something similar, he is merely disowning the embarrassingly naive ontology of palmier days. Unfortunately many of Russell's readers, aided by a dominant philosophical hostility to relations, have taken him to have swept away unities as well as aggregates, thus reducing a series to its members. Since this mistake leads to fatal misunderstandings in wide areas of Russell's philosophy, it is worth emphasising again that this was *not* his view. As he insists,

> . . . we must not look for the definition of order in the nature of the set of terms to be ordered, since one set of terms has many orders. The order lies, not in the *class* of terms, but in a relation among the members of the class, in respect of which some appear as earlier and some as later. The fact that a class may have many orders is due to the fact that there can be many relations holding among the members of one single class.
>
> (*IMP*: 30–1).[19]

Plainly we are here dealing with one of the important properties of relations which has no analogue in the theory of classes and is crucial to philosophy.

Russell suggests a psychological reason why it has been widely believed that the class of its terms suffices to characterise a series. "Sometimes one order is so much more familiar and natural to our thoughts that we are inclined to regard it as *the* order of that set of terms . . . " (*IMP*: 29–30),[20] forgetting that the set is capable of many equally legitimate orderings. Thus the natural numbers in their order of magnitude strike us as being in their 'correct' sequence. When we want to consider other orderings of the natural numbers, we commonly talk about "arranging" them in these orders. But, says Russell, when

> . . . we say that we 'arrange' the numbers in these various orders, that is an inaccurate expression: what we really do is to turn our attention to *certain relations between the natural numbers, which themselves generate such-and-such an arrangement.* We can no more 'arrange' the natural numbers than we can the starry heavens; but just as we may notice among the fixed stars either their order of brightness or their distribution in the sky, so there are various relations among numbers which may be observed, and which give rise to various different orders . . . "
>
> (*IMP*: 30, my italics).

We need, then, in dealing with order, to go beyond the logic of classes to the logic of relations. In *Principia Mathematica* various principles about relations are developed which are important in Russell's later work. Especially noteworthy is "*relation-arithmetic*" by means of which "the concept 'structure' can be precisely defined" (*MPD*: 72, *HK*: 271).[21] Since the term 'structure' figures repeatedly in Russell's subsequent philosophical writings, it is vital to understand its close affinity with relations if we are to do justice to the philosophical views being propounded. Unfortunately for Russell, as will emerge later, this has not been the case. Hence his lament

> . . . that those who are not familiar with mathematical logic find great difficulty in understanding what is meant by 'structure', and, owing to this difficulty, are *apt to go astray in attempting to understand the empirical world.* For this reason, if for no other, I am sorry that the theory of relation-arithmetic has been largely unnoticed.
>
> (*MPD*: 76, my italics).

Just as Russell's philosophy involves relations as well as particulars in its treatment of series or order, likewise relations (or structure) as well as particulars are essential to its treatment of complex arrangements of entities. Hence the central role of structure in an account of the empirical world.

The fundamental notion in relation-arithmetic is that of "similarity of relations" or "likeness" or "ordinal similarity" (*IMP*: 52, *MPD*: 72). In the theory of classes, two classes with the *same* number of terms are said to be *similar*. There is a *one-one correlation* between the two classes because there is a one-one relation whose domain is one class and whose converse domain is the other. Whereas similarity of classes establishes relations between *members* of classes, similarity of relations establishes relations between *relations of members* of classes. For example, if two series have the same number of members ordered respectively by the relations P and Q, then to say that there is an *ordinal similar-*

ity or *likeness* between the relations P and Q is to say that not only is there a one-one correlation between the field of P and the field of Q, but *also* that whenever two terms have the relation P their correlates have the relation Q, and vice versa. In this example, P and Q are serial relations, but

> . . . the conception of likeness is applicable to all relations that have fields – i.e. to all relations where the domain and converse domain are of the same type.
>
> (*MPD*: 73).[22]

Among examples of similarity of relations considered by Russell are the case of precedence among married male government officials and their wives (*MPD*: 73) (where P is the relation of precedence among the officials, Q is the relation of precedence among the wives and the relation of wife to husband establishes the one-one correlation between the fields of P and Q) and the case of the relation between a map and what it maps (*IMP*: 52–3, *HK*: 270) (where spatial relations on a map are correlated with spatial relations at a locality). It is noteworthy that ordinal similarity involves both relations that establish correlations between different classes and relations that generate series, i.e. two of the three sorts of "specially important" relations discussed previously.

The *relation-number* of a relation P is the class of all those relations that are ordinally similar to P. From this basis, the whole of relation-arithmetic can be developed along the same lines as cardinal arithmetic. Note that for Russell relation-numbers, like numbers, reduce to classes, but not so the relations. Most important, from the point of view of this book, is the notion of structure. Relations that are *similar* (or have *likeness*) generate the same structure, i.e. two relations have the same structure when they have the same relation-number. The structure of a relation

> . . . does not depend upon the particular terms that make up the field of the relation. The field may be changed without changing the structure, and the structure may be changed without changing the field. . . .
>
> (*IMP*: 60).

Thus, when "two relations have the same structure, their logical properties are identical, except such as depend upon the membership of their fields" (*MPD*: 75). This means that if the first relation is (say) transitive, asymmetrical and connected, i.e. serial, then so is every relation that has the same structure. Likewise, if the first relation "is one-many or one-one, the other is one-many or one-one, and so on, through all the general properties of relations" (*IMP*: 54–5). So when Russell says that structure "is important for empirical reasons" (*MPD*: 75), he has in mind the principle that the logical structure of experience corresponds to logical structure in the external world. For example, "a gramophone record and the music that it plays are indistinguishable as regards their logical properties although the empirical material of which they are composed is very different in the two cases" (*MPD*: 76). Hence the general Russellian claim that

> . . . what matters in mathematics, and to a very great extent in physical

science, is not the intrinsic nature of our terms, but the logical nature of their interrelations.

(*MPD*: 79).

However, this is to anticipate what comes later. The main point of this chapter has been to establish that analysis and relations go hand-in-hand in mathematical philosophy. Whilst the foregoing discussion of aspects of relation theory has prepared the way for an understanding of Russell's use of relations in general philosophy, it has also provided clear instances of the intimate connection between analysis and relations in mathematical philosophy. Each of the three "specially important" kinds of relation plays a vital part in Russell's programme of analysis, i.e. to deduce mathematics (the *results*) from logic (the *premisses*). Relations that give rise to descriptive functions (one-many relations) underlie many mathematical functions. A certain kind of one-many relations ("selections") underlies multiplication (*IMP*: 117ff.). Relations that establish one-one correlations between classes underlie the "most elementary properties of numbers" (*IMP*: 195–6). Relations that generate series abound in mathematics, order in both number theory and geometry being obvious cases. However analysis shows that other fundamental notions such as *dimensions* in geometry and *limit* are based on serial relations (*IMP*: 29). Even more significantly, though it might seem when starting analysis that mathematics is centrally "the science of number", the movement back to logic (the premisses) via considerations that depend heavily on relation theory, has made it "possible to generalise much that used to be proved only in connection with numbers" (*IMP*: 195). Thus "traditional arithmetic is at once dissolved and enlarged" in "a set of deductive systems" based largely on relation theory, thereby bridging the gap between logic and arithmetic.

Hence, I conclude that the practice of analysis in mathematical philosophy, as outlined in Chapter Two, invoked relation theory in many and varied guises. So much so that the close connection between analysis and relations in Russell's philosophy warrants further investigation.

NOTES

1. All quotations in this paragraph are from *MPD*: 66–7.
2. For discussion of Russell's doctrine of internal relations see Griffin, 1991: 323–5 and 1992: 189–90.
3. In OQ, AMR and AMR/TS. For discussion see Griffin, 1991: Section 8.5.
4. *Principles of Mathematics* and *Principia Mathematica* are, of course, Russell's major contributions to logicism. For detailed discussion of Russell's logicism and its limitations see Grattan-Guinness, 1984 and 1990 and Rodriguez-Consuegra, 1991.
5. See, e.g., PLA: 183–4 and *IMP*: 44–5.
6. The relational fact, aRb, is first construed as a class of ordered pairs, <a, b> ε R. In the second step, the class of ordered pairs is construed as a class of classes, {{a}, {a, b}} ε R for Kuratowski and {{a, Λ}, {b}} ε R for Wiener. These reductions would seem to offend against the theory of types. Perhaps this explains Russell's (and Wittgenstein's) far from flattering view of Wiener's work (Clark, 1975: 215–7).
7. From the point of view of Russell's realism about meanings, the Kuratowski/Wiener defini-

tions as accounts of what relations actually are, or what relational words mean, are incredible. They imply that each trivial rearrangement of the contents of the world changes the meaning of some relational word. (I owe this point to Nicholas Griffin.)

8. Vuillemin's 'platonist' tag clearly overlaps my use of 'Platonist' to describe Russell's 1899–1913 phase. As was shown in Chapter Four my 'Platonist' entails a lot more than the existence of universals, with which Vuillemin's 'platonist' is essentially concerned.
9. See also pp. 124–6 and LA: 164–5.
10. The claim here seems to be that the *ordering* itself is just as much a part of existence as the elements that are ordered.
11. See, e.g., Black (1971).
12. See also *MPD*: 68–9. The material in the next four paragraphs is mainly taken from these two sources.
13. Cf. *IMP*: 43, "The question how to *construct* relations having some useful property by means of operations upon relations which only have rudiments of the property is one of considerable importance."
14. In effect, the relation 'descendant in the direct male line' is substituted for 'son' and the converse domain of this relation is restricted to ancestors of x in the direct male line.
15. Russell's example is the relation of $x+1$ to x, where x is a finite cardinal integer. This relation leads to the descriptive function $(+1)$ 'x; it correlates with any class α of numbers the class obtained by adding 1 to each member of α ; and, finally, the relation can be used to generate the series of natural numbers in order of increasing magnitude.
16. Cf. *PoM*: 242.
17. Cf. *PLA*: 237: ". . . a series is a logical fiction."
18. The *Principles of Mathematics* doctrine of propositions whereby an unasserted proposition is apparently equated with a fact or state of affairs is certainly peculiar since it seems to involve a simple *category mistake*. However the basic point about the analysis of unities (or relational complexes) remains in force when Russell abandons the notion of aggregates as entities. See ERB for a clear statement on this.
19. A similar argument to this one, based on *co-extensive properties*, poses a well-known difficulty for *class nominalism*, i.e. the view that seeks to reduce apparent references to *properties* to references to *classes* of particulars. For details see Armstrong, 1978, Vol. 1: 35.
20. See also *PoM*: 242.
21. Cf. *AMa*: 249, " . . . structure . . . is not applicable to classes, but only to relations or systems of relations."
22. See also *AMa*: 249.

CHAPTER 6

Relations in General Philosophy

6.1 Relations as the Basic Objects of Experience

Earlier chapters have demonstrated the close link between analysis and relations in Russell's mathematical philosophy. Yet, as usually happens in philosophy, developments in one area are not long in spilling over into others. Hence it comes as no surprise to find Russell applying the combination of analysis and relations to a much wider range of philosophical problems. Indeed, for Russell, mathematical philosophy itself extends into the realm of space, matter and motion (*PoM*: Parts VI and VII).[1] His view is that pure mathematics is concerned in an *a priori* way with *possible* spaces, Euclidean and non-Euclidean, and *possible* dynamics, Newtonian and non-Newtonian. This view survives from Russell's idealist phase.[2] Thus the application of the method of analysis and relations to the philosophy of the empirical world is merely a logical continuation of what went before. We shall find that relations are essential to the account of Russellian analysis in general philosophy developed in Chapters Three and Four above. If anything, the union of analysis and relations in general philosophy is even more intimate than in mathematical philosophy. However, we first need to consider some fundamental points in Russell's views about relations.

Of crucial importance to this chapter is Russell's doctrine that we *directly experience* (are acquainted with) relations. However before outlining this thesis in detail, some further aspects of Russell's neo-Hegelian views on relations need to be considered. Firstly, as Griffin (1991: 182) warns us, there "... is a serious danger . . . that [Russell's] early views on relations never formed a coherent theory." This is so because not only are his views spread over works written at different times on various subjects, they are often not very explicit and may be, in some cases, ideas simply taken over from other sources. Certainly the changing role of relations in British idealism, as it evolved under the influence of Hegel, was a complex one. Whereas, for example, Bradley was convinced of the unreality of relations, T.H. Green found a place for them in his idealism.[3] So there was no question of Russell simply agreeing with Bradley on relations in his neo-Hegelian phase, and then later attacking his views from a Platonist perspective.[4] While there are thus some uncertainties about Russell's neo-Hegelian views on relations, some pertinent points can be added to the fact that his adherence to a doctrine of internal relations gave rise to various paradoxes and antinomies (as outlined in the previous chapter).

Except for two brief periods of doubt, Russell was always a pluralist during his neo-Hegelian phase (Griffin, 1991: 80). He also rejected external relations. This raises the issue of how Russell accounted for diversity without external relations. The answer is what Griffin (1991: 327) identifies as Russell's "little noticed theory of proto-relations." 'Proto-relations' is the term suggested by Griffin for things which are not themselves genuine relations, but which are necessary conditions for relations. Examples of proto-relations include identity and externality (Griffin, 1991: 185). Russell's view is that

> . . . [e]very relation . . . involves a diversity between the related terms, but also some unity. Mere diversity does not give a ground for that interaction, and that interdependence, which a relation requires. Mere unity leaves the terms identical, and thus destroys the reference of one to another required for a relation.
>
> (*EFG*: 198).

Thus, according to Russell, identity provides the unity of a relational situation, while externality provides the diversity. Hence both are necessary conditions for relations, without themselves being relations.

Whatever the merits or otherwise of this theory, it can be seen as an obvious precursor to Russell's doctrine that we *directly experience* (are acquainted with) relations, which as already noted, is of crucial importance to this chapter. This doctrine survived intact throughout the major changes and developments in Russell's post-idealist philosophy.[5] Thus in 1910 he held that we are "directly aware" of relational complexes, and of their components, i.e. we perceive *particulars* having *relations* to one another (KAKD: 150). (At this time Russell also held that in addition to our awareness of relational complexes and their components, we are aware of universal relations, i.e. awareness of *concepts* as well as *percepts*. However, this is not important to our present point.) In 1913 Russell defined analysis of "a complex . . . with which we are acquainted" as "the discovery of the constituents and the manner of combination of [the] complex" (*Papers* 7: 119). He added that "the analysis is *complete*" only "when we become acquainted with all the constituents and with their manner of combination." Once again, acquaintance with relations is to the fore. By 1918, he was talking about *facts* as the objects of perception, where particulars having relations to one another constitute one kind of simple fact (PLA: 176–7). From the 1920's onwards, when, influenced by neutral monism, Russell had dropped the notion of acquaintance as an irreducible two-term relation between subject and object, and thereby refined his views on sensation and perception, he nevertheless retained the doctrine of our *direct experience of relations*. Thus in 1946 he wrote:

> What must there be in the world besides qualities, in order that qualities

> may form the pattern in which we find them? Clearly we cannot describe our experience by merely mentioning qualities, we must also take account of relations, for it is in virtue of relations that qualities are arranged in patterns. . . . The relations up-and-down, right-and-left, are *just as much part of what is perceived as are the actual colours* . . . I do not see how we can escape the conclusion that 'above' or 'to the right of', in a suitable context, have as much factual meaning as any words can possibly have.
> (PU: 31, my italics).

The direct experience of relations is, then, a consistently recurring doctrine in Russell.[6] In addition, relations enjoy something of a special position amongst objects of direct experience. This is so for at least two reasons. Firstly, relations, more so than other entities, resisted Russell's various attempts at reduction. Secondly, relations represent the most secure data for our knowledge of the external world. On the first point, particulars and qualities, together with relations, constitute objects of direct experience. (See, e.g., KAKD: 150.) Particulars directly experienced are sense-data (up to 1918) and percepts (1919 onwards) but *not* physical objects (i.e. the 'common sense' objects of direct realism).[7] After an early flirtation with the notion of simple (irreducible) particulars, Russell came to hold that all particulars might, in principle, be complex.[8] That is to say, further analysis, might show how to replace the said particulars by complexes of different particulars related in appropriate ways. In Section 4.1, it was emphasised that when we take account of both the analysis and the synthesis steps in this process, the relations are seen to be an integral part of the complexes. In short, then, there is a trend in Russell to analyse particulars into complexes of *simpler particulars plus relations*. Since these simpler particulars are in turn liable to analysis of the same kind, such analysis has the effect of multiplying tokens of particulars *and* relations, though not necessarily *types*.

In the case of qualities, Russell also offers a reduction, though not with much enthusiasm. The reduction follows the resemblance nominalist strategy of analysing qualities as relations of likeness between particulars. In Russell's hands this is turned into an argument against nominalism on the grounds that, far from dispensing with universals, the analysis merely replaces one sort of universal (qualities) by another (the relation of *likeness*).[9] In 1912, Russell concluded that this argument:

> . . . shows that universal qualities *can* . . . be replaced by exact likenesses, of various kinds between particulars. This view has, so far as I know, nothing to recommend it beyond its logical possibility.
> (RUP: 172).

By the 1940's Russell had developed a new theory of qualities: the notoriously difficult doctrine that qualities such as *red* are particulars whilst *colour* is still a universal. The subtleties of this doctrine need not detain us here (they are discussed in Section 12.4). At best, then, there is a question mark about whether qualities are ultimate.

When it comes to relations, it turns out that Russell never found any satisfactory way of effecting a *general* reduction. Instead reductions were developed for certain limited cases, e.g. it is "formally possible" to replace symmetrical relations by predicates (*IMP*: 44). Likewise, some relations can be *defined* in terms of others (*IMP*: 45). In the heady days of his initial liberation from idealism, it was *asymmetrical* relations ("the most characteristically relational of relations" (*IMP*: 45)), which impressed Russell as the paradigm of irreducible relations. (Which was not to deny that *some* asymmetrical relations were reducible, but only to *other* asymmetrical relations (*PoM*: 256).) Though asymmetrical relations retain their ascendancy in Russell, in the 1940's, he (typically) gives serious consideration to the suggestion that all other relations are reducible to the single *symmetrical* relation of likeness (or similarity). However, at best, Russell is very dubious about this reduction succeeding (*IMT*: 345–6, PU: 33–4).

Hence, I conclude that of the objects of direct experience, relations are most basic in the sense that:

a) reduction of the other objects has a tendency to increase the number of relations, and
b) the most important relations are not reducible except possibly to other relations.

6.2 The Epistemological Primacy of Relations

What of the second reason why relations enjoy a special status amongst objects of direct experience, viz. that relations, and not particulars or qualities, provide our soundest data about the external world? It is a persistent theme in Russell that all we can know of the physical world in general is its *structure*. Structure, and the related notion of *order* (both of which, as we saw in Chapter Five, depend on relations), are the basic elements in Russell's *representative realism*, which is underpinned by the causal theory of perception, and is also the basis of his rejection of Kant's philosophy. As against the Kantian position that the raw material of sensation is the only aspect of experience due in part to the external world, with relations, whether spatial, temporal, or whatever, being supplied by the mind, Russell adopts a contrary view. According to him qualities are subjective, although *externally caused*, whilst the relations between their instances closely represent the structure of the external world, i.e.

> . . . all the actual qualities in percepts are different from those in their unperceived causes, but . . . there is a certain structural similarity between the system of percepts and the system of their causes.
>
> (*HWP*: 688).[10]

Kant's weakness, says Russell, is that he allows

> . . . that the mind orders the raw material of sensation, but never thinks it necessary to say why it orders it as it does and not otherwise.
>
> (*HWP*: 687).

The answer, according to Russell, is that the order in the mind is a more or less exact representation of the order in the external causes of the sensations. The relationship of Russell's philosophy to Kant's will be taken up further in Chapter Nine.

The thesis of the epistemological primacy of relations starts as a tentative psychological claim in 1905:

> . . . we have much more belief in the accuracy of our perceptions of relations than in that of our perceptions of qualities . . . It is only by holding fast to relations as perceived that science manages, on an empirical basis, to construct a world so different from that of perception. Why we should trust in our perception of relations I do not know; but it is a fact that we do so.
>
> (SH: 593).

However, over the years it grows in significance to a universally applicable scientific principle of which human perception is but one instance:

> . . . if A and B are two complex structures and A can cause B, then there must be some degree of identity of structure between A and B. It is because of this principle that a complex of sensations can give us information about the complex that caused them.
>
> (*HK*: 486).

The treatment of structure in mathematical philosophy was obviously not far from Russell's mind as he developed and extended this thesis of the epistemological primacy of relations. This thesis also provides the basis for Russell's solution to the old primary/secondary quality problem, as the discussion in Chapter Nine will show.

As was shown in Chapter Five *order* and *structure* in Russell involve relational realism.[11] So when Russell declares that in "ontology I start by accepting the truth of physics" (RTC: 700), this, combined with the claim that physics is all about structure, puts relations at the top of the agenda. Unfortunately, Russell himself muddied the waters on this point. Given the general philosophical antipathy to relational realism, statements such as "assuming physics true, what we know about the physical world is very little" (*HK*: 345) and

> . . . [i]t is not always realised how exceedingly abstract is the information that theoretical physics has to give. It lays down certain fundamental equations which enable it to deal with the logical structure of events, while leaving it completely unknown what is the intrinsic character of the events that have the structure.
>
> (*MPD*: 13).

are bound to confuse. Since they depend for their plausibility on relational realism, they are liable to appear far-fetched to most readers and thereby lead to rejection of Russell's philosophy.

6.3 Analysis and the Proliferation of Relations

Obviously, the preceding discussion far from exhausts the complex topic of the role of relations in Russell's general philosophy. However, it provides enough material for us to now substantiate the thesis that analysis and relations are just as closely linked, if not more so, in Russell's general philosophy as they were in his mathematical philosophy. Oddly enough, Russell's use of Occam's Razor constitutes a good starting point for the argument. As is well known, Russell was from his Empiricist phase onwards, a staunch advocate for Occam's Razor, "the maxim which inspires all scientific philosophising". According to this maxim "entities are not to be multiplied without necessity" (*OKEW2*: 112), or, in Russell's favourite formulation: "Wherever possible, substitute constructions out of known entities for inferences to unknown entities" (LA: 164)[12]. It will now be argued that failure to understand Russell's use of Occam's Razor has obscured the close links between analysis and relations in his general philosophy.

A very misleading interpretation would go something like this. Occam's Razor is "the principle of ontological economy",[13] i.e. it legislates against multiplying *types* of entities.[14] Since most philosophers find particulars to be less problematic than universals (both monadic and polyadic) or classes, Occam's Razor enjoins them to be nominalist about qualities, relations, classes and the like, if at all possible, and to simply furnish the world with particulars. On this view Russellian analysis requires us to replace inferred particulars (such as Russell's desk) by constructions out of known particulars (sense-data). A constructed particular is a mere collocation *of* simpler particulars. From such a viewpoint, Russell's commitment to universals (in some sense) is seen as a failure, a fatal defect in his philosophy. ("He tried hard to follow Occam's advice, and get rid of them, but unfortunately didn't quite succeed.")

This *ontological* interpretation of Occam's Razor is not Russell's use of the term at all – as will now be shown. Indeed it departs so grossly from Russell's own interpretation that nobody with any familiarity with Russell's writings would think otherwise. As far as I know, nobody has explicitly attributed this interpretation to Russell. However, something much more insidious has happened. So widespread is the ontological interpretation of Occam's Razor, that many philosophers combine that interpretation with features of Russell's own interpretation to produce a version of his philosophy not unlike the one outlined in the previous paragraph.

What are the grounds for holding that the foregoing is a travesty of Russell's use of Occam's Razor? The first point to notice is that neither Russell nor Whitehead adopt an *ontological* interpretation of Occam's Razor – for them it is an *epistemic* principle. This becomes clear from closer inspection of the contexts in which Russell discussed Occam's Razor. Thus in the three places from which the previous quotations are drawn, he adds the following glosses on the principle:

> In other words, in dealing with any subject-matter, find out what entities are undeniably involved, and state everything in terms of these entities. Very

> often the resulting statement is more complicated and difficult than one which, like common sense and most philosophy, assumes hypothetical entities whose existence there is no good reason to believe in.
>
> (*OKEW2*: 112).

> When some set of supposed entities has neat logical properties, it turns out, in a great many instances, that the supposed entities can be replaced by purely logical structures composed of entities which have not such neat properties. In that case, in interpreting a body of propositions hitherto believed to be about the supposed entities, we can substitute the logical structures without altering any of the detail of the body of propositions in question. This is an economy, because entities with neat logical properties are always inferred, and if the propositions in which they occur can be interpreted without making this inference, the ground for the inference fails, and our body of propositions is secured against the need of a doubtful step.
>
> (LA: 164).

> . . . by the principle of Occam's Razor, if the class of appearances will fulfil the purposes for the sake of which the thing was invented by the prehistoric metaphysicians to whom common sense is due, economy demands that we should identify the thing with the class of its appearances. It is not necessary to *deny* a substance or substratum underlying these appearances; it is merely expedient to abstain from asserting this unnecessary entity.
>
> (RSDP: 11).

In each of these quotations the emphasis is clearly *epistemic*: stick to entities that are "*undeniably* involved" such as the "class of *appearances*" of a 'thing' and thereby "our body of propositions is secured against the need of a *doubtful* step". That this is indeed the Russellian interpretation of Occam's Razor is further supported by other references to the matter.

> . . . Occam's Razor . . . to get on in philosophy with the smallest possible apparatus, . . . *it diminishes the risk of error*, because it is not necessary to deny the entities you do not assert, and therefore you run *less risk of error* the fewer entities you assume.
>
> (PLA: 195, my italics).[15]

> . . . Occam's Razor . . . comes in, in practice, in this way: take some science, say physics. You have there a given body of doctrine, a set of propositions . . . you go through, if you are analysing a science like physics, these propositions with a view to finding out what is the *smallest empirical apparatus* – or the smallest apparatus, not necessarily wholly empirical – out of which you can build up these propositions.
>
> (PLA: 235, my italics).

All of this establishes that, for Russell, Occam's Razor is primarily an *epistemic* principle rather than an *ontological* one. The former urges us to eschew *non-experienced* entities whenever possible, whereas the latter enjoins us to minimise the number of *types* in our ontology. It is readily apparent that the two are *not* the same. While experienced particulars, such as sense-data or events, are of the *same* logical type as non-experienced sensibilia or events, Russell's non-experienced particulars that are logical constructions of experienced entities are of a different type. Clearly then, the two principles carve up the furniture of the world differently. It is important to realise that whenever Russell applauds his principle for sweeping away "the useless menagerie of metaphysical monsters" (RSDP: 11), it is an *intra-type* economisation rather than an *inter-type* one which is being commended. Similarly, Russell's version of the principle doesn't insist on a reduction in *absolute numbers* of entities (e.g. a single non-experienced point replaced by a series of sense-data and sensibilia).

Since Russell specifically describes the "heuristic maxim" applied by him and Whitehead as "*a form of* Ockham's razor" (LA: 164, my italics),[16] it seems that they were well aware that their version of the principle was not the usual one. This point is supported by what Whitehead wrote in a 1917 essay, which clearly demonstrates that he too saw his use of the principle as mainly epistemic rather than metaphysical:

> Occam's razor, *Entia non multiplicanda praeter necessitatem*, is not an arbitrary rule based on mere logical elegancy. Nor is its application purely confined to metaphysical speculation. I am ignorant of the precise reason for its metaphysical validity, but its scientific validity is obvious, namely, every use of hypothetical entities diminishes the claim of scientific reasoning to be the necessary outcome of a harmony between thought and sense-presentation. As hypothesis increases, necessity diminishes.
>
> (Whitehead, 1970: 218).

I conclude, then, that unless account is taken of the peculiar way in which Russell uses Occam's Razor, serious misinterpretation of his views is likely. The direction that such misinterpretation is liable to take has already been indicated above in the account that results from attributing to Russell an ontological version of Occam's Razor. What will be the major differences between an account faithful to Russell and the one outlined above? Since Russell is attempting to work with the "smallest empirical apparatus" the various objects of direct experience will be central to his efforts. As shown above, relations are not only objects of direct experience, but in important respects are pre-eminent amongst such objects in that they are (i) most resistant to reduction into other entities, and (ii) the basis of physics which represents the most secure knowledge we have of the world (*MPD*: 13).

The differences between the various accounts are clear from the following diagram:

	Experienced	Non-Experienced
Particulars	1	2
Universals	3	4

Figure 6.1

A typical ontological account of Occam's Razor will lead to boxes 1 and 2 appearing in our theory. (More rarely, boxes 3 and 4.) Russell's account leads to boxes 1 and 3 being the basis of our theory. Those philosophers who unconsciously superimpose an ontological account onto Russell's epistemic one will likely end up invoking box 1 only and thinking that Russell does the same.

The clash between the *usual account* of Occam's Razor and the *Russellian account* is now clear. Most provocatively it could be said that, for Russell, Occam's Razor is a principle of *proliferation of relations*, whereas the standard account opposes relations. (Of course, were it not for the widespread hostility to universals, it would eminently comply with the standard account to have an ontology of nothing but relations. However, most philosophers have overwhelmingly endorsed Aristotle's view that "what is relative is least of all things, a kind of entity or substance" (1928: 1088a22).) If this is correct, it will be clear that any attempt to understand Russell's philosophy which assumes the standard account is liable to be rather wide of its mark.

This opposition in interpretations of Occam's Razor is reminiscent of Russell's own comment on Bergson's presence at the 1911 Aristotelian Society meeting where "On the Relations of Universals and Particulars" was read: "Bergson . . . remarked, with surprise, that I seemed to think it was the existence of particulars, not of universals, that needed proving" (*MPD*: 120). Bergson was right. Not just for 1911 Russell, but throughout his subsequent development, for the sorts of reasons discussed above, Russell moved against the philosophical tide in the matter of the relative status of universals and particulars.

In the project of proving the intimate connection, between analysis and relations in Russell's general philosophy, we have so far demonstrated that he held:

a) that relations are primary objects of direct experience, and

b) that Occam's Razor, "the supreme maxim of scientific philosophising", which enjoins us to substitute *constructions* of directly experienced objects for inferred entities, proliferates relations, i.e. relations assume a pivotal role in our account of the world.

To clinch the argument it remains to show

c) that Occam's Razor is just a special statement of Russell's method of analysis.

That this is so is fairly readily demonstrated.

In the example, quoted above (PLA: 235) of the application of Occam's Razor to physics so as to isolate "the smallest empirical apparatus", (perhaps with some non-empirical augmentation), out of which to build up the body of propositions constituting that subject, Russell specifically calls it an *analysis*. Likewise two other examples of analysis previously discussed are specifically given by Russell as examples of the application of Occam's Razor, viz. the reduction of cardinal numbers to classes (RSDP: 11) and the reduction of the desk to sense-data (PLA: 236–7). The connection between Occam's Razor and the method of analysis is simply that in analysis of propositions about the external world (e.g. those of physics), we should seek *empirical* propositions as the premisses.

In earlier chapters we found that analysis in its first stage involved a backwards movement from common knowledge that is relatively vague, concrete and complex (the *results*) to more simple, abstract and precise statements which are also less obvious (the *premisses*). The premisses then serve as the basis of a deductive system which displays the original inexact knowledge in a form that is much less vague and approximate. Two characteristics of the premisses, as elucidated in Chapter Three, are especially relevant here. Firstly, the premisses are typically "much less obvious" than the results. Secondly, not all of the premisses are equally certain and, of course, this degree of certainty of the respective premisses carries through the deduction thereby determining the degree of certainty of the results.[17] Part of successful analysis, then, is to seek premisses with a maximal degree of certainty.

At first sight, the fact that premisses are far from obvious might seem to clash with their certainty being maximal. Our discussion of the objects of direct experience shows that this is not so. It *is* far from *obvious* that relations, properties and particulars (not those of common sense) are the objects of direct experience, yet it is equally clear from the above discussion that the premisses for empirical knowledge, according to Russell, should be about these objects.[18] Hence Occam's Razor, as practised by Russell, fits his method of analysis exactly. Therefore the general conclusion of this chapter is established, viz. that the practice of analysis in general philosophy proliferates instances of known relations, i.e. analysis and relations have an especially intimate connection in Russell's philosophy. This applies to both general and mathematical philosophy. The moral is that we should start thinking in terms of *relations* whenever we encounter "series", "structure",[19] "correlation" or the like in Russell's writings.

Someone who has followed the argument so far in this book might nevertheless object as follows. While there is no doubt that Russell makes much of *relations* in his philosophy, whilst taking a *realist* view of them, how important is this in the end? Can't we view relations in a nominalist way without greatly affecting the overall shape of Russell's philosophy? The task of the next chapter is to refute firmly this suggestion.

Notes

1. For discussion of Russell's views on mathematics and their relation to philosophy see Grattan-Guinness, 1976, 1984 and 1990.

2. Early versions of this underlie *EFG*, APG and LOG.
3. Hylton (1990, Part 1) provides an illuminating comparison of Bradley and Green on relations. Green's critiques of empiricism and of Kant centred on their inability to account satisfactorily for the phenomenon of relatedness.
4. For major differences between the doctrines of internal relations in Bradley and the neo-Hegelian Russell see Griffin, 1991: 325–6.
5. In RTC: 693, Russell comments as follows: "Mr Black must suppose me to hold that we cannot be acquainted with relations – a view which I have repeatedly repudiated with all possible emphasis."
6. Other clear-cut statements of the doctrine include *PP*: 101–2; *AMa*: 145; LE: 140; *IMT*: 344 and RTC: 688.
7. Cf. the analysis diagrams for Russell's various phases outlined in Chapter Four.
8. As was noted above in Chapter Four.
9. Russell refers to this argument frequently, e.g. RUP: 172; *PP*: 95–6; PLA: 183 and *IMT*: 346–7.
10. See also *PP*: 85ff.
11. By *relational realism* I simply mean the thesis that *some* relations are an irreducible part of our ontology. Armstrong (1978, Vol. 2: 80ff.) uses the term *relational realism* for the stronger thesis that "every property is analysable without remainder in terms of relations". Armstrong suggests that C.S. Peirce may have been a relational realist in this strong sense.
12. See RSDP: 11 for a very similar formulation of the principle.
13. See, e.g., Flew, 1979: 236.
14. 'Types' here refers to *ontological* types or categories. Types as *universals* are, of course, a sub-set of these.
15. See also PLA: 243; MMD: 14 and RTC: 708.
16. In MMD: 14, Russell describes his use of Occam's Razor as "quite independent of metaphysics."
17. See especially *OKEW2*: 215.
18. So the *epistemological primacy of relations* doesn't mean that relations are the subject of primitive epistemological propositions as described in Chapter Four.
19. For an interpretation of Russell that makes structure *linguistic*, rather than ontological, see S. Hampshire in Magee, 1971: 20.

CHAPTER 7

Logical Constructions and Relations

In Chapters Three and Four the *narrow* and *broad* interpretations of Russellian analysis were distinguished and it was shown that the former stops at the stage of analysis into parts, whereas the latter moves on to the *synthesis* of the parts into a complex structure. It was argued that Russell himself supported the broad interpretation which emphasises the place of relations in the structure, whereas the narrow interpretation tended to overlook relations. Having now seen that relations are central to Russellian analysis of our knowledge of the empirical world, we are in a position to appreciate the consequences of not taking relations seriously. Logical constructions constitute a prime example of the confusion produced in the secondary Russellian literature by such an omission.

7.1 The Different Kinds of Logical Constructions

According to Russell, "logical constructions" or "symbolic fictions" are "[c]lasses or series of particulars, collected together on account of some property which makes it convenient to be able to speak of them as wholes" (UCM: 76–7).[1] It is worthy of note that he says *classes* or *series* of particulars, thereby distinguishing cases where serial *relations* may not be (and typically aren't) involved, from cases where they constitute an essential part of the construction. However, this vital point has generally gone unnoticed. Unfortunately for Russell, the two most famous examples of his wielding Occam's Razor so as to substitute "logical constructions . . . for inferred entities" (RSDP: 11) fall under the *class* case rather than the *series* one. The famous examples are, of course, *cardinal numbers* and *definite descriptions*, and neither typically involves serial relations in its substitute logical construction. (The class of all pairs, for example, is the substitute for the cardinal number two; a class of existential statements is the substitute for the definite description.) Such is the fame of these class cases that when Russell turns to the example of a physical object, such as his desk, being a logical construction out of sense-data (early Russell) or out of events (later Russell), people take him to mean that it is a collection or class of particulars (with relations having no essential part in the construction). Yet this is precisely one of those logical constructions that is a *series*, as distinct from a class, of particulars, i.e. its particulars fall under a serial relation. It is an indication of how misunderstood are relations in Russell's philosophy that when he defines logical constructions as "classes or series of particulars", many think the two are interchangeable. In fact, the secondary Russellian literature abounds with examples of writers who think *either* that all logical constructions fall under the *class* case *or* (what amounts to the same thing) that the *series* case is not significantly different from the class one. Thus Sainsbury asserts that "all Russell's constructions are classes" (1979: 239). Jager believes that in *Our Knowledge of the External World*, "an external object is identified with the total class of all the actual and possible sense-data associated with it" (1972: 293). Lewis agrees: "The thing is . . . to be defined as the totality (or 'class') of all its possible aspects. . . . " (1968: 55).[2] Of course, in the series case the construction is still a *class*, but crucially and more importantly, it is something else as well. It is this something else that is repeatedly overlooked.

As demonstrated above, series involve *order*, i.e. relations, as well as elements, whereas mere classes involve only elements. However, there are quite a number of reasons why this simple distinction (and its wider importance for Russell's philosophy) has been missed. For a start there is the complication that on occasion Russell himself *appears* to suggest that ordinary physical objects are *classes* of sense-data rather than series of the same. This apparent equivocation on Russell's part is dissipated by taking into account two vital distinctions: (a) an object *at an instant* as against an object *over time*, and (b) the viewpoint of a *single* observer (i.e. a *perspective*) as against the viewpoints of many observers, both actual and possible (i.e. *a series of perspectives*). This pair of distinctions generates the four cases set out in the following table.

Table 7.1

A physical object	At an instant	Over time
For a single observer	A class of sense-data (visual, auditory, tactile, etc.)	A series of classes of sense-data
For many observers	A series of classes of sense-data	A series of series of classes of sense-data

For a *single observer at an instant,* what common sense calls a physical object presents itself as a *class of sense-data*[3] each member having a relation of compresence to one another, but no serial relation. For *a single observer over time*, the various momentary classes of sense-data are found to have *serial relations* to one another as well as the temporal relation of succession. The serial character of the successive sense-data is readily apparent in cases where there is relative motion between the observer and the common sense physical object. Take for example, a steam train leaving an observer at the station. At each moment there is a class of visual, auditory, tactile and olfactory sense-data, which together form a series of classes over time. Obviously there are serial relationships between the various classes (the visual size of the train progressively decreases, as does its sound, the shaking of the platform and the smell of the smoke). In addition there are correlations between the changes *within* sense-data kinds (as visual size halves, sound intensity is quartered (say); below a certain visual size there is no sound, etc.). This example features straightforward decreasing serial relations. Russell generalises the idea of such serial relationships between successive classes to all cases, no matter how complex they may be.[4] (Instead of describing the case of the single observer over time as a *series of classes* of sense-data, it seems equally satisfactory to describe it as a *class of series* of sense-data, where there is a series of visual sense-data, a series of auditory sense-data, etc. with definite correlations between members of the respective series.)[5] The case of a physical object presenting itself to *many observers at an instant*[6] has obvious affinities with the previous one. Rather than each member of the series of classes of sense-data belonging to a single observer at different times, each class in the series consists of the combined sense-data of many different observers *all* at the very *same instant*. (The series is to be thought of as including the sense-data of actual and possible observers.) The close correspondence between the two cases is clear when we compare the series of classes of sense-data of a single observer successively observing this desk from *n* different positions, with the series of classes of sense-data of *n* observers, each at one of the *n* different positions, observing the desk at the same instant. It is obvious that the serial relations between the classes are *similar* in the two cases, even though according to Russell, they wouldn't be *identical* because sense-data are physical particulars that only last a very short time and hence two people can't experience *exactly* the same sense-data.[7] If this is so, our single observer moving around the desk would experience successively slightly different sense-data from those experienced simultaneously by the *n* observers.

Finally there is the case of the physical object presenting itself to *many observers over time*. Whereas the previous case involved series in three dimensions at an instant, the present case *extends* these series of classes over time,[8] and so represents even more complex arrangements of sense-data. In common sense terms not only is there a gradual change in the appearance of the desk viewed from different angles, but there is a similar gradual change in appearance over time when viewed from the same position. This last case is, of course, Russell's full reconstruction of a physical object in the normal sense of that term. The other cases are merely important stages in the exposition of his sense-data theory of physical objects.

In terms of *logical constructions* then, only a physical object for a *single observer at an instant* falls under the *class* analysis. Each of the other three cases involves ordering-relations and therefore is a logical construction of the *series* kind. When Russell himself *appears* to suggest that physical objects are mere classes of sense-data, there are, as far as I can see, two likely explanations. Either, (a) he is discussing the single observer at an instant case (i.e. physical objects in the most tenuous sense) or, (b) he is using "class (or set) of sense-data" as a convenient abbreviated way of referring to the complex series of series of classes of sense-data that constitute physical objects in their fullest sense. I hasten to add that in the latter case there is no question of a change of doctrine because, typically, the discussion makes it very clear that serial relations are part of the construction.[9] After all, a complex series of series of classes of sense-data is a class of sense-data (it just happens to be much more as well!).

Whilst Russell's tendency to abbreviation in this case is understandable, it is also particularly unfortunate when combined with the misunderstandings about series and classes being both *logical fictions* (see Chapter Five). Someone who views series and classes on the same footing as *mere* fictions, will interpret the move from "series of classes" to "classes" as further confirmation of the unimportance of all the talk about series. For this reason, it is worth stressing once again that the series is a logical fiction in that it is not *another thing* in the same sense as the things that are its members. As argued above, Russell is keen to emphasise this to counter his earlier *Principles of Mathematics* theory to the contrary – a theory that was effectively demolished by developments in the theory of types. But we must not let this obscure the fact that the series is something more than its class of members, this extra something being a pattern or complex relation of a different *type* from the elements it relates.

In summary, classes are ontologically different from series because they consist of their elements and nothing else, whereas series consist of elements having definite relational patterns. So the tendency to see all logical constructions as classes plays havoc with Russell's philosophy. Even the most sympathetic commentators haven't been immune from this lapse, e.g. Weitz has Russell positing "events in the form of 'electrons' and 'protons'" (1944: 74). "Electrons and protons are not events, according to my theory; they are *series of groups* of events" (RTC: 685, my italics)[10] is Russell's entirely appropriate reply.

Evidence that Russell's occasional talk of *classes* when he strictly means *series of classes* is *not* to be understood as a doctrinal retreat is provided by his discussions of the logical construction of spatial points, along lines suggested by Whitehead. For instance in his Herbert Spencer Lecture of November 1914, Russell has Whitehead's method construing "a physical point as a certain class of those objects which are the ultimate constituents of the physical world . . . as a class of physical entities . . . " (SMP: 68–9).[11] Yet in the Lowell Lectures given in March and April of the *same* year, in a detailed discussion of this topic, Russell defines a point as "all the objects which enclose members of a given punctual enclosure-series" (*OKEW1*: 115),[12] where the definitions of the enclosure relation and of a punctual enclosure-series ensure that the objects constituting the point are themselves a series. In other words, the *technical* definition of a point constructs it as a *series*, rather than as a *class*, of objects. This situation is repeated in Whitehead's own writings on the subject. In non-technical terms, a "point is the class of extended objects which, in ordinary language, contain that point" (Whitehead, 1970: 177). (Whitehead goes on to say that "rather elaborate" technical steps are required to remove the obvious circularity in this ordinary language definition.) However, in his more technical moments, Whitehead's definitions of points also invoke relations and series, rather than mere classes. Thus "a point is merely a cumbrous way of directing attention to a set of relations between a certain set of thought-objects of perception . . . " (1970: 206) and a "point in space is nothing else than a certain set of relations between spatial extensions" (1970: 236). So it seems that the urge to explain technical matters in everyday terms is largely behind Russell's talk of *classes* when strictly speaking he means *series of* classes. This in turn has encouraged the notion that all logical constructions are mere classes.

7.2 Logical Constructions and Russellian Analysis

Russell's equivocation about classes and series of classes is not the only factor causing misunderstanding of the theory of logical constructions. A further major reason why the role of relations in logical constructions is generally overlooked stems from doubts about the status of such constructions. Are logical constructions mere symbolic devices or do they have wider ontological implications? At first sight it is not easy to decide this question either way, because Russell can be read as holding either view or as running the two together. When we note that 'logical construction', 'symbolic fiction', 'logical fiction' and 'incomplete symbol' are employed apparently interchangeably by Russell, we would naturally incline to the view that logical constructions are essentially symbolic devices.[13] However, there is also a clear tendency to discuss logical constructions in *ontological* terms: e.g. "classes or series of particulars collected together on account of some property" (UCM: 77) or "construct a logical fiction . . . composed of empirically given things . . . " (PLA: 236).[14] Given that Russell appears to support both alternatives in the same writings, the obvious *prima facie* conclusion is that he is simply confused or careless on the matter.

In order to show that things are not as bad as this, we need to return to our account of Russellian analysis given above. In the process we will resolve the difficulties about the interpretation of the theory of logical constructions and find further reasons why the role of relations in such constructions is overlooked. As we found, the following diagram (Figure 7.1) is a schematic representation of Russellian analysis.

Our reason for returning to the analysis diagram is that it clarifies the links between Russellian analysis in general and some well-known instances of it, such as the method of logical constructions and the theory of descriptions. Whereas other philosophers have tended to concentrate on the special features of each instance, thereby overlooking what the instances have in common, Russell has repeatedly emphasised the latter. An excellent example of this is provided by the theory of descriptions. Despite the recent powerful contrary views of Cocchiarella and Griffin (discussed in Section 4.3.2), it is by now an apparently unassailable plank of 'Russellian folklore' amongst philosophers that it was the theory of descriptions that provided the first instance in Russell's work of the analysis of propositions resulting in a reduction in one's ontological commitments. This view stresses how the theory of descriptions showed that the super-

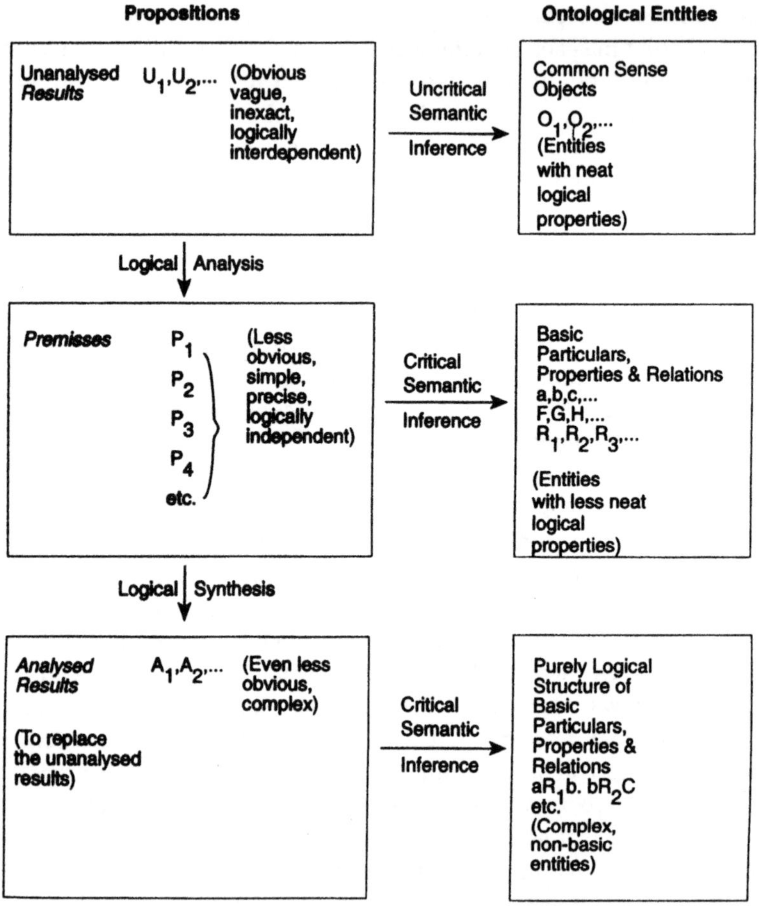

Figure 7.1. Russellian Philosophical Analysis.

ficial grammatical form of a sentence is often quite different from the logical form, so that what appear to be referring expressions at the grammatical level cease to refer at the logical level. Then, so the story goes, Russell extended this 1905 discovery into a whole metaphysical theory about "logical constructions" (or "logical fictions") whereby apparent grammatical reference to a whole range of supposed entities (points, instants, common sense physical objects, etc.) disappeared on analysis, thereby eliminating any need to postulate such entities. Thus the theory of descriptions is seen as the origin of Occam's Razor in Russell's philosophy. The proponents of this view contrast the restricted ontology resulting from the theory of logical constructions with its more luxuriant pre-1905 counterpart, notoriously expressed in *The Principles of Mathematics*:

> Whatever may be an object of thought, or may occur in any true or false proposition, or can be counted as *one*, I call a *term* . . . every term has being, *i.e. is* in some sense. A man, a moment, a number, a class, a relation, a chimaera, or anything else that can be mentioned, is sure to be a term; and

to deny that such and such a thing is a term must always be false.
(*PoM*: 43).

This view, that there was a new direction in Russell's philosophy after 1905 as the method of logical constructions was developed out of general features of the theory of descriptions, has its obvious attractions. After all, the theory of descriptions is usually recognised as one of Russell's seminal contributions to philosophy, so one might expect it to lead to new directions in related areas of philosophy. However, as attractive as this view might be, it is nevertheless false, as was argued in Section 4.3.2. Russell himself emphatically rejects this view:

> My first applications of the method of logical construction were in pure mathematics: the definitions of cardinals, ordinals and real numbers, and the construction of points in a projective space as pencils in a descriptive space. *All these antedated the theory of descriptions*, and were dictated by dislike of postulation where it can be avoided.
> (RTC: 692, my italics).

Russell's claim here is certainly correct. Not only were the constructions he mentions developed *before* 1905,[15] but much of the other discussion in *The Principles of Mathematics* is surprisingly *constructivist* in outlook. Consider what he says about material points, for instance:

> . . . it is plain that the only relevant function of a material point is to establish a correlation between all moments of time and some points of space, and that this correlation is many-one. So soon as the correlation is given, the actual material point ceases to have any importance. Thus we may replace a material point by a many-one relation whose domain is a certain one-dimensional series, and whose converse domain is contained in a certain three-dimensional series.
> (*PoM*: 468).[16]

(In addition, this is an excellent illustration of the thesis that analysis has the effect of proliferating relations!)

So the popular view that the theory of descriptions inspired the *method* of logical constructions is false. What *does* seem to be true, however, is that Russell only used the expression "logical construction" after 1905 and later applied it to his earlier work on the legitimate grounds that it accurately represented some of that early work. Indeed this Russellian habit of introducing new technical terms with seemingly almost every major book or article, is largely responsible for the trend amongst critics to focus on logical constructions *per se* or incomplete symbols *per se* or logical fictions *per se* or . . . In the process it is often overlooked that the same *sort* of thing is discussed elsewhere using different terminology. Russell himself had no such trouble, often taking the contrary course of applying a new technical term to appropriate earlier work. We have just seen that "logical construction" is one clear case of this. Another one is his 1924 survey of the applications of Occam's Razor in his work (LA: 164–8). Instances cited are the principle of abstraction, cardinal numbers, classes, definite descriptions, series, ordinal numbers, real numbers, points, instants, event-particles, matter and the psychological subject. All are said to be "examples of the substitution of constructions for inferences" (LA: 166), something that is only supposed to have started in 1905.

We were led to return to the analysis diagram in order to show that Russell's plethora of related technical terms is as versatile as his own practice assumes. Before doing so we need to attend to something that is likely puzzling the reader. If Russell in *Principles of Mathematics* holds that "every term has being" then how could he also be "substituting logical constructions for inferred entities"? The inferred entities would at least subsist in their own right. A.J. Ayer has noticed this problem and points out that as well as holding that "every term has being", Russell further added that " ... every term is immutable and indestructible. What a term is, it is, and no change can be conceived in it which would not destroy its identity and make it another term" (*PoM*: 44). Ayer explores the suggestion that what Russell has in mind here is that every sign has a determinate Fregean sense (1971: 17–20). He has no difficulty in showing that this leads to absurdity and concludes that what Russell really meant was that in cases like number there is an implicit immutable sense hidden from the public eye. Most people go through a lifetime of thinking of numbers without knowing that they are actually thinking of classes of classes. However, this explanation by Ayer will not do, if only because it leads to the conclusion that the sense of a sign is indeterminate, not just for the uninformed, but for *everyone*. This is so because, as has been demonstrated in previous chapters, Russellian analysis *is unlikely to be final*, i.e. a class of classes is hardly a likely candidate for the "immutable and indestructible" sense of talk about a number, if only because classes are themselves logical fictions, though their exact analysis is somewhat uncertain.[17] Thus Ayer's account has Russell's famous *Principles of Mathematics* doctrine, that "every term has being" and that "anything that can be mentioned is sure to be a term", applying to the never attained ideal of the logically perfect language.

For a more plausible account we need look no further than the Cocchiarella (1987, Ch. 3) interpretation of the *Principles of Mathematics* theory of terms. As was shown in Section 4.3.2, Russell's pre-1905 ontology may not have been as lush as has been generally assumed. In particular, Cocchiarella and Griffin have argued that, for Russell at this time, not all denoting concepts denote. Rather, many of them merely purport to denote. This interpretation enables us to see how Russell could hold in *Principles of Mathematics* that "every term has being" yet also be "substituting logical constructions for inferred entities". In the case of cardinal numbers, for example, these numbers are inferred entities logically constructed as sets of sets. However, while cardinal numbers themselves are thereby reduced, and hence cannot be subsistent entities, for each cardinal number, there is a denoting concept which does subsist (and which is easily confused with the number). Nevertheless, the crucial point is that these are examples of denoting concepts that fail to denote.

In the light of our rejection of the myth that the theory of descriptions provided Russell's initial venture in ontological reduction, it is worth stressing that

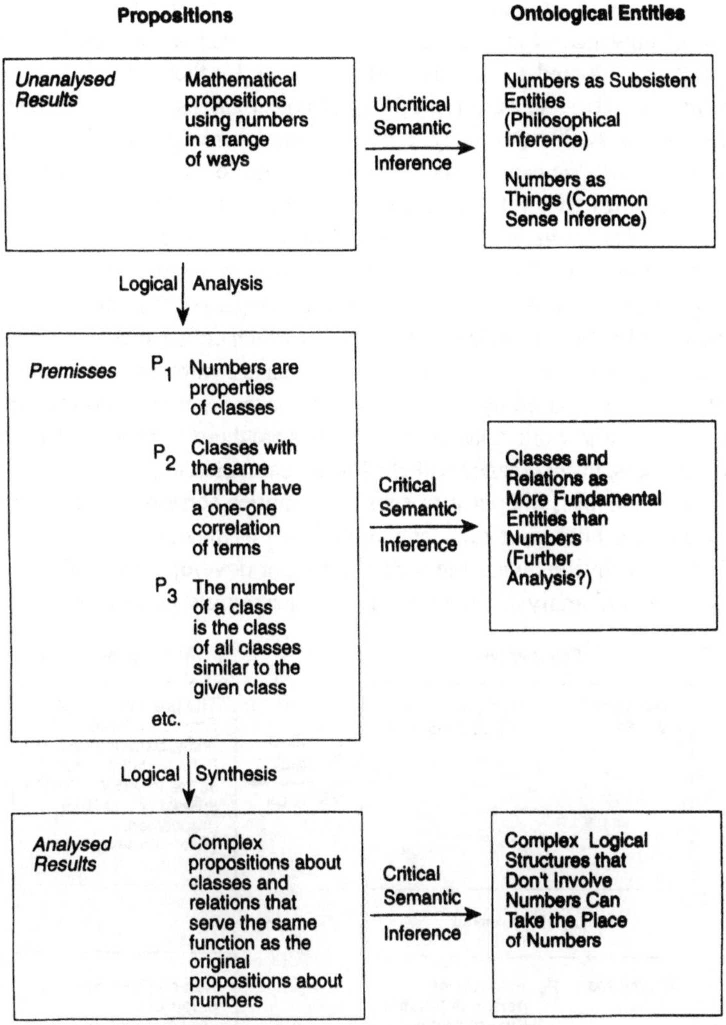

Figure 7.2. Analysis of Cardinal Numbers.

the theory of descriptions was an attack on *subsistent*, rather than *existent*, entities, in particular on denoting concepts (Griffin, 1993a: 175ff.). As a consequence of the theory of descriptions, there is a truncation of, but *not* elimination of, the category of subsistent being.[18] So, for example, the theory of descriptions analyses the proposition "Scott is the author of Waverley" as "The propositional function '*x wrote Waverley* is equivalent to *x is Scott*' is true for all values of *x*". This removes the previously felt need to accord *subsistent* being to "the author of Waverley" in addition to the being of Scott. This sort of reduction in the category of subsistent being occurred in 1905. A few years later, in *Principia Mathematica*, subsistent being was further pruned as classes were shown to be incomplete symbols. "Thus classes . . . are merely symbolic or linguistic conveniences, not genuine objects . . . " (*PM*, Vol. 1: 72). As noted already, however, the category of subsistent being survived at least as late as 1913 in Russell as the home of universals. Later (certainly by 1918) Russell had dropped the notion of subsistence and adopted a single level of existence. He thereafter had no hesitation in applying the term "substitution of logical constructions for inferred entities" to all of his previous efforts at analysis whether they were concerned with subsistent or existent objects. We can use the analysis diagram to clarify these developments.

Early examples of analysis in the philosophy of mathematics, such as the definitions of cardinals, ordinals and real numbers, have analysis diagrams as shown above (Figure 7.2) (taking, e.g., the case of cardinal numbers).

The diagram demonstrates, in non-mathematical language, how analysis of propositions about cardinal numbers leads to premisses which define such numbers in terms of classes, relations and other logical ideas. Logical synthesis enables the original propositions to be replaced by a set of propositions that don't mention cardinal numbers but do the same work. Classes and relations are thereby more fundamental than cardinal numbers. In accordance with the liberal realm of being countenanced in *Principles of Mathematics*, inference to subsistent entities corresponding to terms (e.g. denoting concepts) occurs at each level of the analysis diagram.

Definite descriptions were the next significant development in Russell's philosophy. A typical analysis diagram for this case would be as follows:

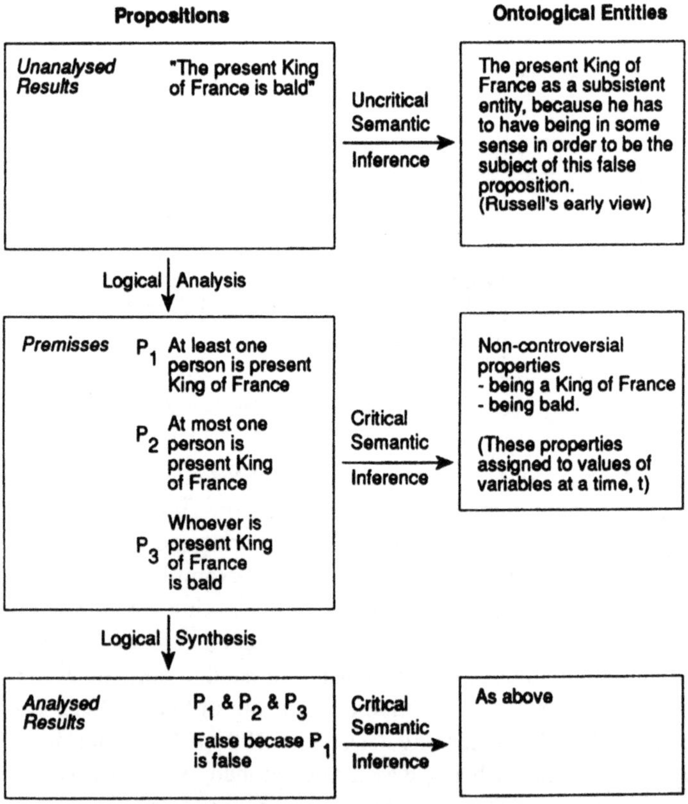

Figure 7.3. Analysis of a definite description.

It is noteworthy that in this case the uncritical inference to a subsistent *particular* is replaced by critical inferences to subsistent *properties*. From 1905 onwards, Russell used such analyses to avoid, firstly, some subsistent entities of his earlier Platonism, such as denoting concepts, then, later, subsistent classes. However, properties and relations remained, even after the notion of subsistence was dropped, subject always to the proviso of further analysis. As argued above, relations proved to be especially resistant to reductive analysis. Russell's conclusion is that when a definite description "occurs in a proposition, there is no constituent of that proposition corresponding to that description as a whole. In the true analysis of the proposition, the description is broken up and disappears" (PLA: 216–7). Definite descriptions are also termed *incomplete symbols* by Russell. Incomplete symbols "have absolutely no meaning whatsoever in isolation but merely acquire a meaning in a context" (PLA: 221).

As already noted, Russell went on to analyse classes as incomplete symbols. The analysis diagram is:

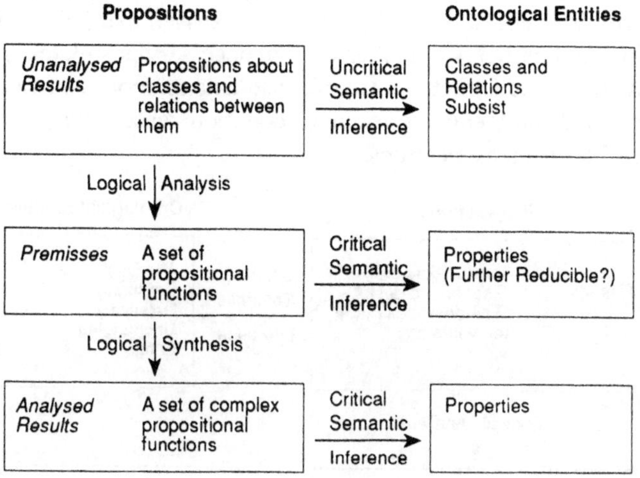

Figure 7.4. Analysis of classes

This, of course, carries the analysis of cardinal numbers a stage further. The *premisses* in our cardinal numbers analysis diagram become the *unanalysed results* on which the new stage of analysis is carried out. Hence cardinal numbers also become "merely symbolic or linguistic conveniences". To take an example from Russell,

> . . . the proposition '1+ 1 = 2', somewhat simplified, becomes the following: "Form the propositional function '*a* is not *b*, and whatever *x* may be, *x is a* γ is always equivalent to *x is a or x is b*'; form also the propositional function '*a* is a γ, and, whatever *x* may be, *x is a* γ *but is not a* is always equivalent to *x is b*'. Then, whatever γ may be, the assertion that one of these propositional functions is not always false (for different values of *a* and *b*) is equivalent to the assertion that the other is not always false."

> Here the numbers 1 and 2 have entirely disappeared and a similar analysis can be applied to any arithmetical proposition.
>
> (*PoM*: x).[19]

The analysis of classes diagram, together with the example from Russell, makes it look as though analysis of mathematics, when taken far enough, dispenses with *any* inferences to ontological entities. This is a mistaken conclusion. As we found in Chapter Five, central parts of mathematics feature propositions about relations which, according to Russell, are definitely not reducible to propositions about classes. Ordinal numbers, for instance, are defined as classes of serial relations, the serial relations being untouched by the analysis of classes (*PoM*: 242–3).[20] Likewise, pure geometry turns out to be "the study of series of two or more dimensions" (*PoM*: 372). So, given our earlier stress on the nature of serial relations, as well as the ability of relations to survive analysis, there will still be ontological entities inferred from the analysis of mathematics.

Next we turn to the analysis diagram for *logical constructions* as Russell first used the term. Whitehead originally persuaded Russell that points of space, instants of time and particles of matter were unnecessary postulates and could be replaced by logical constructions out of sense-data or, more generally, events.[21] Russell extended the technique to cover common sense physical objects and scientific entities such as electrons.

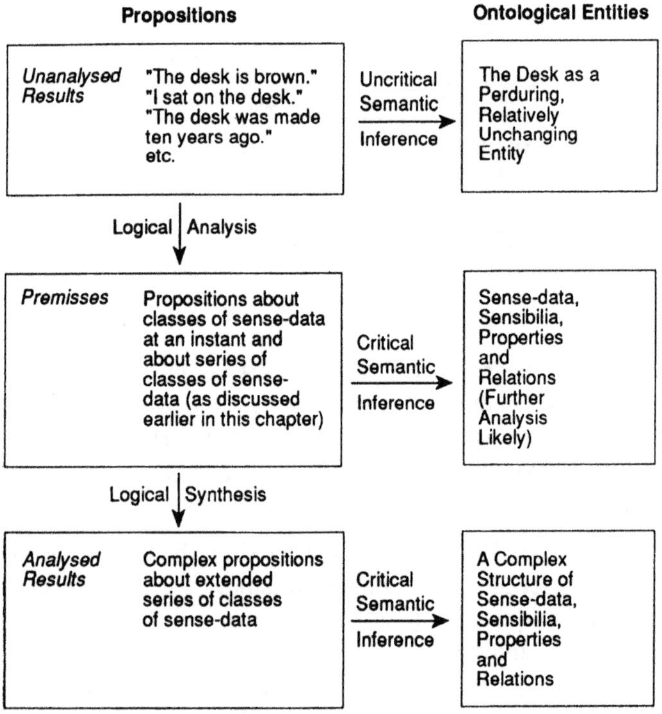

Figure 7.5. Analysis of a common sense physical object.

The overall pattern of this analysis diagram is the same for spatial points, temporal instants, material particles or scientific entities.

I conclude from the foregoing analysis diagrams that the account of Russellian analysis developed above does indeed cover each of the cases considered. In each instance an initial set of propositions is refined by analysis to produce a set of more fundamental propositions. Logical synthesis then leads to a set of novel propositions which replace the initial set, thus completing both stages of the philosophical analysis. Ontological inferences, if any, are made from the set of more fundamental propositions, always with the proviso that further analysis might lead to a revision of such ontological commitments. Hence the *riskier* ontological inferences from the initial set of propositions are replaced by the *safer* inferences to complexes of the more basic ontological units. It follows from this that Russell's tendency to use the terms "logical construction", "incomplete symbol", "logical fiction", etc., interchangeably is harmless.

Likewise, when we remember that analysis is *directly* carried out on propositions and only *indirectly* on entities, his carelessness about the distinction between propositions and the entities they imply is also seen to be harmless. For instance, Russell is fond of saying that "classes . . . are incomplete symbols" (e.g. PLA: 228) when strictly speaking this is an abbreviated (and misleading) way of saying that "the symbol for a class is an incomplete symbol; it does not really stand for part of the propositions in which symbolically it occurs, but in the right analysis of those propositions that symbol has been broken up and disappeared" (PLA: 228). So when Russell occasionally says that *entities* are incomplete symbols, he merely means that the *symbols* for these entities are expunged by analysis. Similar remarks apply to the term "logical fiction". It is clear that for Russell, logical fictions are *entities*, whether real or apparent, rather than *propositions* about entities (PLA: 234, 236–7). However, he occasionally seems to hold otherwise, e.g., logical fictions are said to be "the same thing" as "aggregations of symbols" (PLA: 221).[22] In this case, the context makes it clear that he means that the aggregations are incomplete symbols, while the logical fictions are the entities to which the symbols apparently refer. "Symbolic fiction" and "logical construction" are similar to "logical fiction" in that they primarily apply to *entities* though they are occasionally applied carelessly to *symbols* for entities. The claim is that the analysis diagram clarifies this by making Russell's intentions absolutely clear.

We have been discussing the confusion that can arise in regard to Russell's technical terminology. We found that "incomplete symbol" belongs to the left hand column of the analysis diagram, whilst "logical fiction", "symbolic fiction" and "logical construction" apply to the right hand column. Unfortunately besides the *horizontal* confusions which the diagram clarifies, there are also *vertical* confusions. Whilst *incomplete symbols* obviously occupy the top of the left hand column and are replaced by *other* symbols at the bottom of the column, the situation is not so clear-cut in the right hand column. When Russell spells out what he means by "logical construction", "logical fiction", etc., it is clear that they belong at the *bottom* of the right hand column and are substituted *in our ontology* for the entities at the *top* of the right hand column. However, Russell is also fond of saying that supposed entities, such as Socrates, Piccadilly, the desk (which belong at the top of the right hand column) *are* logical constructions or logical fictions.[23] As the analysis diagram shows, to say that the table is a logical construction is merely a shorthand way of saying that the inferred entity, the table, is expunged from our ontology and replaced by a logical construction out of more basic entities. So, strictly speaking, the table isn't accorded existence, but in another sense it is, in that there *is* a complex entity which has certain of the features that common sense attributed to the table.

The moral of all this is that we need to be very wary of jumping to facile conclusions about these matters on the basis of what Russell says in one or two places in his writings. This applies especially to such notions as "logical construction" which we have found to be subject to (in terms of our analysis diagram) *both* horizontal and vertical ambiguity. Confusion can be avoided by adhering to certain basic principles that can be extracted from the analysis diagram:

1. *Construction* occurs *vertically* (never horizontally or diagonally), i.e. analysed results (propositions) are constructed out of premisses (basic propositions) and complex entities are constructed out of basic entities (such as events, properties and relations). The precise sense in which entities are 'constructed' will be investigated shortly.
2. *Substitution* of constructions occurs *vertically* (never horizontally or diagonally), i.e. analysed results (propositions) are substituted for unanalysed results (propositions) and complex, constructed entities are substituted for simple, common sense, inferred entities.
3. *Inference* is only ever horizontal (never vertical or diagonal).
4. Analysis, as was noted previously, occurs in the left hand column, i.e. on propositions. (Entities are only analysed indirectly.)

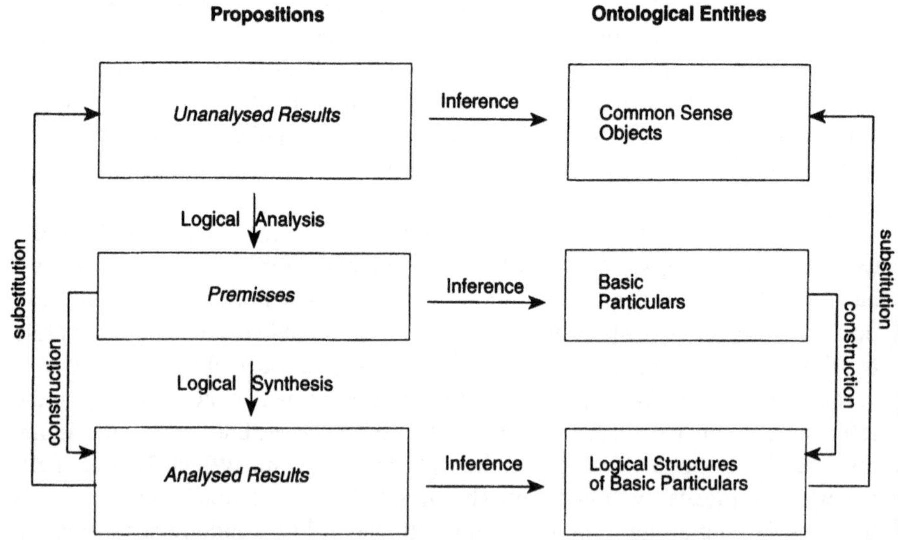

Figure 7.6. Basic features of Russellian Philosophical Analysis.

The analysis diagram reveals the complexity and subtlety of Russellian analysis. This is important if we are to avoid hasty criticism of Russell's views based on simplistic understanding of what he was trying to do. That this need is real is shown, e.g., by the major study of Russell's philosophy in which Sainsbury claims that "the method of analysis . . . involves reducing complex truths to simpler truths, and complex things to simpler things" (Sainsbury, 1979: 235). By taking this scanty (and misleading) characterisation of the first stage of analysis as the whole story, Sainsbury is able to reject Russell's construction of common sense physical objects on the grounds that the "overall effect of construction is to increase, rather than diminish, the number of inferred entities" (1979: 202). Russell is thereby convicted of sinning against Occam's Razor, his own "supreme maxim in scientific philosophising" – and Sainsbury is not even aware that the constructions involve relations to further swell the numbers of entities! As our account has demonstrated, Russell never promised a "gross tonnage" reduction of entities, while the inferred entities that are replaced stem from *riskier* inferences than those involved in arriving at the entities that do the replacing.[24] In short, logical constructions out of more soundly assumed entities replace inferences to less soundly assumed entities. Of course *directly known* entities are the paradigm of sound ontological assumptions, but Russell never relied *solely* on experienced basic entities as the detailed discussion in Chapter Four showed.

7.3 Misinterpretations of Russell's Logical Constructions

Because of the doubts engendered by Russell's writings on the question of whether logical constructions are mere symbolic devices or whether, instead, they also have ontological import, discussions of the question tend to come down decisively on one side or the other. In both cases, the role of relations in logical constructions is missed. We have seen above how logical constructions are concerned with both symbolism and what is symbolised, with relations playing a central part in our account. Those who emphasise the merely symbolic side of logical constructions usually do so on the grounds that they add nothing to the furniture of the world, i.e., on analysis propositions making apparent reference to certain entities are replaced by propositions which make no such reference; so the entities are logical constructions.[25] In support of such a view, Fritz, for example, argues that a

> . . . 'construction' is normally something actually built by some rearrangement of material . . . but in Russell's 'constructions' no such building takes place. In the normal sense of the word they are not constructions at all. There is no new object produced as a result of the construction, in fact, just the reverse is the case. The material used in the 'construction', e.g. particulars, has previously existed, and continues unchanged, but is merely classified in a new and distinctive manner.
>
> (Fritz, 1952: 214).

Thus, for Fritz, "a construction is not a *thing*; it is not an entity actually composed of particulars, since this is one type of entity that constructions are designed to make unnecessary" (1952: 217–8). Likewise a logical construction "is not identical with a class of particulars, since this class in turn is an incomplete symbol, and defined by means of a construction" (1952: 218). (Fritz commits himself to the *class* interpretation of logical constructions outlined above, thereby overlooking the *series* interpretation and the relations that they imply.) From considerations such as these Fritz concludes that a logical construction is "an actual procedure of analysis" of a term "which shows that the term is, in fact, an incomplete symbol and can be given a definition in use".

Such a conclusion represents a very narrow account of logical constructions and forces Fritz to hold that Russell is "not correct" in his usage when he seems to say that material objects, for instance, are entities constructed out of other entities (1952: 218). Similarly, he denies that Russell ever *meant* that logical constructions, as *entities,* could be substituted for other entities, despite Russell's explicit statements to the contrary (1952: 216). Our analysis diagram account above has shown that construction *is part* of Russell's understanding of analysis, and to that extent Fritz is correct, but the diagram also shows us that construction and substitution of *entities* are central parts of Russell's overall philosophical analysis. Fritz, and those like him who restrict logical constructions to the realm of symbolism, are required, rather implausibly, to dismiss as mistaken half of what Russell says on the subject.

The reason why people adopt such a view, I now argue, ultimately arises out of a failure to appreciate Russell's doctrine of relations. When Fritz asserts that logical constructions "are not constructions at all" because there is "no new object produced", rather the "material used . . . has previously existed, and continues unchanged, but is merely classified in a new and distinctive manner", he is essentially correct. However, when *relations* are included in the "material used" a different story emerges from the one told by Fritz. Adopting a nominalist, *class* interpretation of logical constructions leaves us with only particulars and classes as the material to be used in construction. On this interpretation, constructions are mere classifications of already existing particulars and classes, as well as their properties and relations understood in a nominalist way. The construction can be *represented* as a truth functional combination of propositions about the various particulars, but the particulars themselves are untouched. Hence the claim that logical constructions are merely symbolic. There is no sense at all on this account, whereby a new object is correctly said to be created. At most the construction might involve so-far-unsensed particulars (sensibilia) which are there whether or not anyone happens to sense them.

In contrast, a Russellian *series* interpretation of logical constructions allows us *relations*, as well as particulars, for the material used in construction. Once again there is no question of construction in the grossest sense. On this point, it is worth recalling a comment by Russell, discussed in Chapter Five, about arranging numbers in an order:

> . . . we say that we 'arrange' the numbers in these various orders, that is an

> inaccurate expression. . . . We can no more 'arrange' the natural numbers than we can the starry heavens; but just as we may notice among the fixed stars either their order of brightness or their distribution in the sky, so there are various relations among numbers which may be observed, and which give rise to various different orders . . .
>
> (*IMP*: 30).

So including relations, as well as particulars, will not enable us to create *new* arrangements of particulars. Fritz is correct that the "material used . . . has previously existed, and continues unchanged" – the world continues as before. However, the relations still make an important difference. Because they are *universals*, there *is* a sense in which their inclusion in logical constructions can introduce "a new object" into the world. Of course there is no suggestion of *new* relations (or particulars) being *produced* by constructions. However logical construction may draw to our attention something that has previously gone unnoticed, viz. that a certain set of particulars falls under a complex relational structure. In this way, we discover new *instances* of relations. When this is combined with the doctrine that the complex relational structure is more than the set of its particulars, i.e. that the relations are part of the ontology of the state of affairs, it is not so unreasonable to talk of a new object. Since a certain set of objects may be well known without anyone suspecting that they have certain complex relations to one another (e.g. forming a complicated series), it is surely accurate to talk in terms of *logical construction* and *substitution* for the process whereby it is pointed out that:

a) the set of objects is united by the complex relational structure; and
b) such a structure has all of the characteristics of some inferred or unknown object, and hence dispenses with the need for postulating such an object.

I conclude that Russell's use of the term 'construction' is not as perverse as is often claimed by those who concentrate on the symbolic aspects of logical construction. The term is certainly subject to the ambiguities previously clarified by the analysis diagram, but nevertheless there is a clear sense in which construction *does* take place. (More will be said on this shortly in our discussion of the other account of logical construction.) In the meantime, logical construction is seen to be another instance of the principle that Russellian analysis tends to proliferate relations.

Turning, on the other hand, to those who accept that logical constructions *are* ontological entities[26] (including those who combine this with a recognition of their role as symbolic devices) we find that the importance of relations is again overlooked. This is so because the *class* interpretation of logical constructions prevails. The favoured analogy is to view the collocations of sense-data or events along the lines of bricks or building blocks in a wall.[27] While Russell himself once or twice talks like this,[28] it is more significant that he expressly, and repeatedly, repudiates this way of understanding logical constructions. "The particulars are to be conceived, not on the analogy of bricks in a building, but rather on the analogy of notes in a symphony" (UCM: 77).[29] The significance of this repudiation is precisely that the brick analogy has the effect of playing down relations whilst the symphony analogy has exactly the opposite effect. Expanding on this Russell says that the

> . . . ultimate constituents of a symphony (apart from relations) are the notes, each of which lasts only for a very short time. We may collect together all the notes played by one instrument: these may be regarded as the analogues of the successive particulars which common sense would regard as successive states of one 'thing'. But the 'thing' ought to be regarded as no more 'real' or 'substantial' than, for example, the role of the trombone. (UCM: 77–8).

If one approaches the matter with a nominalist outlook, it appears that Russell's reason for preferring the symphony analogy to the brick one is just that musical note instances are short-lived (like Russellian sense-data and events), whereas bricks are relatively perduring and unchanging. Hence, it might be concluded that as long as one remembers that the bricks are short-lived, it is harmless to stick to the more familiar building block analogy for logical constructions. I take it that such a line of thought explains the peculiar reluctance on the part of interpreters and critics to take more seriously the musical analogy which Russell himself emphatically prefers.

However, this line of thought, far from being harmless, supports a fatally defective account of Russellian logical constructions. This is because, as Russell himself points out in the above quotation, the ultimate constituents of a symphony are *both* the notes and their *relations*. Given the ontological significance of relations for Russell – a point that we have already repeatedly emphasised – the symphony analogy turns out to be very suitable for underlining this significance in a way that the building block analogy misses. A number of considerations point to these conclusions:

(I) There are some obvious reasons why the nominalist dismissal of relations sits more easily with the building block analogy than with the symphony one:

(a) *Variety of relations*:

The bricks that make up a building are typically of one (or at most, a few) kinds, whereas the number of kinds of musical notes is much greater. This means that the number of *kinds* of relations is very much greater in the symphony case.

(b) *Complexity of relations*:

The much greater variety of relations between musical notes is further enormously increased by the fact that different notes can be sounded simultaneously (chords, etc.), whereas bricks are laid one after another.

(c) *Significance of relations for the whole*:

Most of the relations between individual bricks are irrelevant to the whole represented by the finished building, e.g. it makes little or no difference to a given wall whether each brick overlaps two halves of the bricks in the course below it or one-third of one and two-thirds of the other. In either case the pattern of bricks and relations is repeated to produce a wall of the required size. This is not the case with a symphony. Slight changes in the relations between notes make all the difference between a good performance and a poor one (*HK*: 479), i.e. the whole depends on the relations between the notes being the correct ones. Unlike the wall, the repetition in a symphony is likely to be evident in the particular notes (*n* soundings of middle C, etc.), more so than in the relations between them (symphonic development depending on subtle changes in the relations as the work unfolds).

These three factors show that while relations are comparatively unimportant in the building block analogy, they are of overwhelming importance in the symphony one.

(II) Unlike a wall, a symphony is characterised by its *structure* (in the Russellian sense of that term). This is shown by the fact that a symphony (or other musical composition) can be transposed and/or arranged for various groups of instruments. Thus the piano transcription of Beethoven's Fifth Symphony is still the Fifth Symphony because, despite some of its ultimate constituents being altered or removed (a piano instance of middle C instead of twenty first violins in unison playing middle C or the piano's inability to match the orchestral quality and colour of (say) a passage for the brass), an *overall structure* is maintained in the original and the transcription. The complex relational structures in the two cases are similar.[30] In the case of a wall, the *structure* (i.e. the overall complex of relations between the parts) is not at all crucial, e.g. the same wall requirement can be met using bricks of a different size and shape producing, therefore, a different structure. That the structure, rather than the particular notes, is what characterises a symphony is further shown by the possibility of two symphonies (e.g. by Haydn) consisting of exactly the same *class* of note instances, yet being totally dissimilar as wholes. Such a range of variability is not found in the possible walls that can be formed by a given class of bricks. Even if, as seems likely, the difference between the two cases is merely one of degree rather than of *kind*, the limitations of the building block analogy for representing Russell's position are, by now, only too obvious.

(III)

> When some set of supposed entities has neat logical properties, it turns out, in a great many instances, that the supposed entities can be replaced by purely logical structures composed of entities which have not such neat properties.
>
> (LA: 164).

As this well known statement emphasises, it is a mark of logical constructions that they lack neat logical properties. So the infamous desk, for instance, will be replaced by a logical construction in which

> . . . there shall be no assumption of a metaphysical substance which is identical throughout. It is a little easier to the untrained mind to conceive of an identity than it is to conceive of *a system of correlated particulars, hung one to another by relations of similarity and continuous change and so on.* That idea is apparently *more complicated*, but that is what is empirically given in the real world, and substance, in the sense of something which is continuously identical in the same desk, is not given to you.
>
> (PLA: 237, my italics).

Now the problem with the building block analogy is that walls have the neat logical properties that we are supposed to avoid. Even if we allow considerable variability in the size and shape of the bricks in the wall they still have to fit into the overall pattern. Russell warns us to be "suspicious of anything so tidy and exact" (LA: 167). A symphony, on the other hand, has a complexity that requires training and experience to unravel, so that a

> . . . person without musical training, if he hears a symphony, acquires a vague general impression of a whole, whereas the conductor, as you may see from his gestures, is hearing a total which he minutely analyses into its several parts.
>
> (*MPD*: 169).

So the conductor's minute analysis reveals a complex detailed structure (proliferates relations), which the untrained casual listener misses. In contrast, a wall is a poor analogy for a logical construction because, although a bricklayer might see more than the layperson, there is not a whole lot more to see. This is clear when we note that any reasonable sized part of a typical wall has basically the same structure as any other similarly sized part of the same wall. Obviously this is not so for the various parts of a symphony. These intuitions could be quantified, of course, by means of relation-numbers, as defined in Chapter Five above.

I conclude from all this that Russell's favoured analogy for logical constructions serves not only to emphasise that his *particulars* are short-lived, but also that the *relations* between the particulars are an essential part of the construction. (This applies, of course, to constructions out of sense-data or events – we have seen that there are constructions which don't require such raw material, e.g. classes.) The building block analogy is misleading because relations play an insignificant part in the analogy, whereas the symphony analogy is appropriate because it pushes relations into the prominence they occupy in Russell's overall philosophy. Since most Anglo-American philosophers both in Russell's time, and since, have favoured the nominalist overtones of the former analogy, it is hardly surprising that his ontological understanding of relations, and their role in logical constructions, has been generally overlooked.

The symphony analogy is also helpful on the vexed question of 'In what sense, if any, do *we do* the constructing?' As we saw, there are those who say that only symbols can be constructed, the world that they symbolise being left as it is. Against this it was suggested that drawing attention to previously unknown relations between a set of particulars *is* a sort of ontological construction, especially if, like Russell, you hold a realist theory of relations. This position is supported by the symphony analogy. If a symphony is a logical construction, according to Russell, then in what sense does its composer construct it? It seems that a Russellian account requires us to say that rather than *creating* new relations, the composer *makes us aware* of previously unknown complex relations that hold between the musical notes that comprise the symphony. Just as we can't arrange the stars in the heavens or the natural numbers, though we can call attention to previously unnoticed relations between them, likewise the imaginative composer discovers novel relations between musical notes and, through performance of his/her work, draws them to public attention. So when a musicologist claims that in his Symphony Numbers 3 and 7, Sibelius astonishes us by revealing hitherto unsuspected possibilities of C Major, he is talking in a way that Russell would commend. Hence, when we include relations in our account, it is no more controversial to say that a philosopher logically constructs an electron, than it is to say that a composer constructs a symphony. As is so often the case, neglect of relations is the cause of an ongoing pointless dispute about Russell's philosophy, in this case, about whether 'construction' was the best word for him to use.

So the symphony analogy has major advantages over the building block one. Nevertheless, like all analogies, it also has its limitations. The crucial point is that a logical construction is a created abstract entity which will serve the same function as some inferred entity, e.g., in explaining data. The logical construction is more secure epistemically because all of the elements of the construction are objects of acquaintance. So, while the logical construction is something new, its elements are types of particulars and relations previously known by acquaintance. The result is a proliferation of tokens of familiar entities.

The difference that results in our interpretation of Russell's philosophy when we recognise the role of relations in logical constructions is, then, a matter of some importance. It is worth pointing out that the inclusion of relations rebuts an argument that has been widely regarded as fatal to Russell's theory of logical constructions. This argument is due to Ayer[31] and claims that the theory is "very ingenious", but nevertheless *circular*. The ground for the charge of circularity is that there seems to be no way of picking out from the multitude of sense-data just those from which a common sense object, e.g. a penny, is to be constructed, without reference to the penny itself. Ayer's strategy is to consider in turn possible ways in which such a construction might be based on sense-data alone and, finding fault with them all, thereby conclude that the penny needs to be invoked in its own construction. Ayer first considers the possibility of selecting the sense-data for the construction by their *qualities*, but thinks this is clearly unsatisfactory because "different pennies look very much alike", so "the only way in which we can hope to avoid mixing up sensibilia which belong to different things is by bringing in the other constituents of the perspectives in which they occur" (1971: 62). This latter suggestion involves a consideration of the *relations* between penny sense-data and *other sorts of sense-data* in the same perspective. Ayer rejects this second method of selecting sense-data for the construction on the grounds that two pennies could be placed on separate pieces of the same coloured cloth in such a way that either piece of cloth occupied the whole of a person' s visual field. The argument is that "there will be nothing to show that the two perspectives do not belong to the same set" (1971: 63). Sainsbury thinks that Russell's theory is in just as much trouble if the two pennies are placed against *different* backgrounds. Seizing on Russell's claim that a sense-datum is not the whole sensory field at one time, but "rather such a part of the whole as might be singled out by attention" (RSDP: 6), Sainsbury concludes that sense-data are not distinguishable as belonging to different pennies by their relational properties, since such properties are irrelevant. So, Sainsbury claims, Russell's constructions involve the absurdity that "sensibilia which are intrinsically indistinguishable but associated with distinct pennies will be gathered . . . into a single 'penny'" (1979: 248).

Ayer's response to his "two penny" objection is to propose a third method of selecting sense-data on the basis of how the perspective which they occupy is related to other perspectives. This will involve the determination of *adjacent perspectives* in terms of overlap of common constituents. However, Ayer thinks that this undertaking is beyond the resources of Russell's theory on a number of grounds, the main ones being that

a) we can't determine whether unperceived perspectives have constituents in common; and
b) Russell's rule for determining the spatial ordering of perspectives is too simplistic from both the viewpoints of physics and physiology.

Hence, Ayer concludes, Russell's theory fails.

On the contrary, I claim that the *series* interpretation of logical constructions evades both Ayer's argument and its extension by Sainsbury. A minor initial point would be that Russell's theory is *not* circular if it mentions the penny as an object for any reason other than "convenience of exposition".[32] Russell is not rejecting the penny theory outright, since he accepts many of its predictions about sense experience. He is, instead, pointing out that the theory goes too far in postulating a perduring object that lies behind the sense-data, when a *series* of sense-data serves the same end. So there is no harm in talking about the penny in those areas in which the two theories are in agreement.

However the important, and decisive, point against Ayer and Sainsbury is that they don't consider *relations between the penny sense-data* as the criterion for inclusion in the construction. (Ayer comes closest by considering relations between perspectives, but doesn't think they can be determined.) As demonstrated earlier in this chapter, the penny is, for Russell, an extended (in both space and time) *series of classes of sense-data*. The "two penny" objection (in both the Ayer and Sainsbury forms) depends on considering merely some penny sense-data *at an instant* (and their relations to their sensory surroundings at that instant), whereas Russell's theory is concerned with related classes of sense-data over time. These relations are serial in character, e.g. if (in common sense terms) a penny is moved from the table to the mantelpiece above a roaring fire, then (in sense-data terms) there is a class of sense-data at t_0 (penny visual sense-datum against table-top sense-datum, cold metal tactual sense-datum, etc.) and another class of sense-data at t_n (penny visual sense-datum against mantelpiece sense-datum, warm metal tactual sense-datum, etc.) There are classes of sense-data at $t_1, t_2, t_3, \ldots, t_{n-1}$ which have determinate serial relations to these two classes of sense-data. So with the two pennies on the same kinds of cloth, it would be their respective prior histories that determined that they are indeed two, and not one. In each case, it is a matter of the same kind of inductive generalisation. The Ayer-Sainsbury arguments depend for their plausibility on being restricted to sense-data *at an instant*. Once we consider *series of sense-data over time* their indistinguishability arguments are no longer compelling. As for Ayer's claim that the laws of perspective are more complex, from both a physical and physiological viewpoint, than Russell allows, this is not fatal to Russell's position, even if correct. All that it would entail would be that the series of sense-data was more complicated than Russell thought. However, a series over time, however complicated, can still be inductively established thereby defeating alleged indistinguishability cases.

A further point of interest concerns Ayer's other objection to the third method of selecting sense-data, that we can't tell whether unperceived perspectives have similarities with other perspectives. Certainly, from a nominalist viewpoint, it is a serious difficulty if certain particulars needed in a construction are missing. (Just as a brick wall is liable to be rather unsatisfactory, if too many of its individual bricks go missing!) However, if we adopt the alternative realist viewpoint that takes the relations between the particulars as seriously as the particulars themselves, then the situation is less drastic. If a series has missing members there are still relations between the remaining members that go some way towards preventing a complete discontinuity. (Just as a performance of a symphony is merely a bad or incomplete one, rather than a mere performance of some parts, when a few notes here and there are missing.) The point, simply, is that relational structures have features that nominalists overlook.

This point has ramifications well beyond Ayer's argument, as can be seen by briefly considering a general argument against phenomenalism put forward by J.J.C. Smart. According to Smart phenomenalism faces the difficulty that in replacing common sense objects by groups of sense-data "it is a huge accident that they hang together in the way that they do" (1963: 25). Thus if we take the example of a cat standing behind a large flower pot so that (in common sense terms) we can see her head on one side of the pot and her tail on the other, Smart proposes a problem for phenomenalism as follows. Whereas common sense has a ready explanation of the head part of the cat being aligned with the tail part, viz. that the middle part of the cat is behind the pot, for phenomenalism it is "a cosmic coincidence", a mere "brute fact" that the head sense-datum is aligned with the tail one (Smart, 1977: 24). Since Russell's logical constructions out of sense-data are often seen as a variety of phenomenalism, it is of interest that the *series* interpretation can overcome Smart's difficulty. The overall *relational* structure of a cat is *not* drastically affected by the middle part being missing, i.e. the *relations* between the head and the tail remain even when the 'middle member of the series' is absent. The 'brute' fact that a cat as a common sense object has a certain shape, is matched by the 'brute' fact of the kind of serial relation that underlies Russell's logical construction of the cat. Both 'brute' facts are open to further explanation.

I take it, then, that this chapter has demonstrated decisively that in the many examples of logical constructions that include serial relations and the like, Russell's position will be badly misinterpreted if the relations are not treated in a realist way.

NOTES

1. Cf. PLA: 221: "... what I call 'logical fictions' ... embrace practically all the familiar objects of daily life.... Most of them are *either classes*, or *series*, or *series of classes*" (my italics). See also *MPD*: 101.
2. See also Broad, 1957: 59; Dorward, 1951: 29; Fritz, 1952, especially Ch. IV; Clack, 1969: 66.
3. *OKEW2*: 94 and 117, and *AMi*: 126 are very explicit on this point. PLA: 170–1, implies the same thing.
4. The 'single observer over time' case is explicitly dealt with in UCM: 86; PLA: 238; and *AMi*: 126.
5. See *AMi*: 127.
6. This case is explicit in *OKEW2*: 96 and 117 and *AMi*: 126
7. RSDP: 8–10, 13 and *OKEW2*: 109.
8. This case is explicit in *OKEW2*: 111–2 and 115–6. See also PLA: 170; UCM: 86; *AMi*: 126 and RTC: 706.
9. There are clear instances of (b) in Russell, e.g. early in RSDP physical objects are *classes of appearances* (pp. 10, 14), yet when the discussion becomes detailed they are *series of appearances* (pp. 20–2). See also *AMi*: 98ff. and *HK*: 289.
10. See also *AMa*: 319ff.
11. See also LA: 166.
12. The relation of enclosure is empirical, i.e. it "is given in sense-experience". (See Weitz, 1944: 109.) In his 1926 revision of *OKEW*, Russell removed the paragraph containing this quote because he came to view Whitehead's construction as unsatisfactory for reasons given in *AMa*: 291–4.
13. E.g. "logical constructions or symbolic fictions" (UCM: 76); logical constructions are clearly equivalent to logical fictions in PLA: 235–7, while logical fictions are equated with incomplete symbols in the same work (p. 221).
14. See also LA: 168 "... logical constructions, like all other constructions, require materials ...".
15. *PoM*: 115ff.; 242ff.; 149ff. and 270ff.; and 394ff.
16. See also the discussion of motion, *PoM*: 469–73.
17. For discussion see Quine, 1972.
18. Russell himself points this out very clearly in his 1937 Introduction to the 2nd Edition of *Principles of Mathematics*. He describes (*PoM*: x) the theory of descriptions as *the first step* away from the *Principles of Mathematics* doctrine that "every term has being".
19. The 1937 Introduction to the Second Edition.
20. Russell says (p. 242) that the "correct analysis of ordinals has been prevented hitherto by the prevailing prejudice against relations".
21. See, e.g., *OKEW2*: 7–8; SMP: 69; RSDP: 12; *PoM*, 1937 Introduction to the Second Edition: xi and *MPD*: 77.
22. This point is noted by Brown, 1978: 44.
23. See, e.g., PLA: 171 and 241.
24. The useful term "gross tonnage" was suggested by Keith Campbell of the University of Sydney.
25. Amongst those taking such a line are Clack (1969) and Fritz (1952).
26. See, e.g., Sainsbury, 1979: 200ff; Jager, 1972: 295ff; and Dorward, 1951: 26ff.
27. See, e.g., Sainsbury, 1979: 6, 26, etc.
28. LA: 167, "... the bricks out of which the world is built ... are events ...".
29. See also *ABCR*: 155, 158; *HK*, Part VI, Ch. VI; *MPD*: 169.
30. Though not in the strict Russellian sense defined in Chapter Five, e.g. the piano transcription lacks some of the notes that are in the orchestral score of the symphony. Russell (*AMa*: 227) has musical notes forming series, with all that that implies about relations between them.
31. Ayer, 1971: 62–3. Sainsbury, for instance, says that Ayer's argument shows that Russell's programme of construction has "a glaring and irremediable error at its very first stage" (1979: 248).
32. Contrary to Ayer's claim, 1971: 62.

CHAPTER 8

The Distinctiveness of Russellian Analysis

If the central findings of the preceding chapters are correct, then there is a more subtle structure in Russellian analysis than has usually been recognised. An understanding of this structure should greatly assist in the settling of disagreements in Russellian scholarship. For instance, we have found that Russellian analysis is primarily analysis of *propositions* and only indirectly is it analysis of *entities*. Yet equally clearly, we have found that Russell's main interest is in the latter part of the enterprise. This is enough to show that Ayer's disagreement with Weitz, noted in Chapter One, is of little consequence. Weitz is quite correct in his finding that it is only by analysis of symbols that Russell is able to analyse what they symbolise. Equally, Ayer is correct that Russell's prime interest is in the "ontological question". What remains at issue between Ayer and Weitz is the merely verbal question of whether the "central part" of Russellian analysis is ontological.

The distinctive features of Russellian analysis that have emerged in the previous chapters have largely gone unnoticed. This is particularly true of the *synthesis* stage of analysis and its tendency to proliferate relations. Certainly there has been an awareness of the *importance* of analysis to Russell's overall philosophy. Weitz's classic paper was discussed in Chapter One. As far as it goes, I do not think that it has been bettered by subsequent authors. Eames has rightly asserted that "to understand fully the method of analysis as Russell employs it is to understand his philosophy" (1969: 56). She addresses much of what Russell has to say about analysis, but her view that it "issues in a situation in which . . . simples have replaced complexes . . . " inevitably means that the synthesis step and the role of relations are overlooked. The same deficiency is found in Jager's (1972) study of Russell's philosophical development. Pears seems to have been the first to employ diagrams to clarify Russellian analysis. In his detailed study of the influence of the British empiricist tradition on the development of Russell's philosophy from 1905 to 1919, Pears uses analysis diagrams to illustrate Russell's theories of meaning and truth.[1] These analysis diagrams are somewhat different from the ones used in this book, since they are concerned with the *causal* links between words (both singular and general) and things. What is missing from the Pears work is, once again, recognition of the importance of the synthesis stage of analysis and the significance of relations in that stage. Since the role of *relations*, unlike analysis, has been a rather neglected topic in

Russellian scholarship, it is no surprise that its significance for *analysis* has been generally overlooked.

Why hasn't more attention been paid to the role of relations in Russell's philosophy? The answer, I suspect, springs from Russell's own coyness about discussing relations after the end of the Platonist phase. This coyness is due to the fact that, while the "question of relations is one of the most important that arise in philosophy, as most other issues turn on it . . . ", the usual ways of talking about relations "are inconsistent with the doctrine of types" (LA: 169–70). So in his later works, Russell is much more wary of talk about relations as entities, preferring to stop at relational facts:

> I do not see how we can escape the conclusion that 'above' or 'to the right of', in a suitable context, have as much factual meaning as any words can possibly have.
>
> (PU: 31).

> I think it is as certain as anything can be that there are relational facts such as 'A is earlier than B' . . . There certainly are complex wholes which have a structure, and we cannot describe the structure without relation-words.
>
> (*MPD*: 128).

> If it is true, as it seems to be, that the world cannot be described without the use of the word 'similar' or some equivalent, that seems to imply something about the world, though I do not know exactly what. This is the sense in which I still believe in universals.
>
> (RTC: 688).

Because all words "are of the same logical type" (LA: 169), Russell is able to avoid transgressions against the doctrine of types, in the three preceding quotations, by talking about *relation-words* rather than *relations*. Russell's position here is that although 'above' has, in a suitable context, "as much factual meaning" as any word "can possibly have", nevertheless "it is by no means clear that it means anything in isolation" (PU: 31).[2] Hence, "'above' is dependent upon its terms in a way in which they are not dependent upon it" (PU: 32). So, in response to

> . . . the question: 'What feature of the world is it that makes it necessary to use relation-words in describing it?' . . . we must not answer by saying 'there are relations' but we may answer by saying 'things are related'. Even this, however, will need careful interpretation; it cannot mean 'between the two things A and B some relation holds', for the phrase 'some relation', on the face of it, requires us to treat a relation as a substantive.
>
> (PU: 33).

It is considerations such as these that lie behind Russell's later insistence that, for the problem of universals, the "solution must be technical, and dependent upon considerations derived from modern logic . . . " (PU: 21). He admits that this "does not answer the metaphysical question . . . ", adding that ". . . it comes as near to giving an answer as I know how to come" (*MPD*: 128). Clearly, none of this represents a swing to *nominalism* about universals. Rather, the urgent need is to find a *new* way to formulate *realism*. The crucial point remains that for Russell throughout his pluralism, the *relatedness* aspect of relational facts is as ontologically significant as the particulars that are related.

Of course, it must be admitted that in the preceding chapters, I have ignored transgressions against the doctrine of types in discussions of relations and universals in general. This is no more than Russell himself usually does when he is not directly addressing the problem of universals. The excuse in both cases is that the essential points would remain, even though their expression would be much more cumbersome if conforming to the doctrine of types. Nevertheless, it might be objected that in the analysis diagrams of preceding chapters, I have depicted critical semantic inferences from premisses to relations as *entities*. My response is that this is part of my ignoring type problems, whilst admitting that Russell is actually less certain of the exact ontological implications of relational facts, as the discussion of the last few pages has shown. However, this issue assumes major significance only if one supports a narrow interpretation of Russellian analysis – an interpretation that I have already shown to be mistaken.

To see why this is so, recall that in Chapter Four it was shown that the question of whether analysis had reached ultimate simples or not assumed major importance only for the narrow interpretation of analysis. For Russell, with his adherence to the *broad interpretation*, it was an issue of much less significance. The same sorts of considerations apply to the current issue of contention. If, as the narrow interpretation maintains, analysis reduces complex things to simple things, then Russell will have to specify the entities that constitute a relational fact in order to analyse it. On the other hand, if analysis also includes a *synthesis* step, as the broad interpretation maintains, then analysis typically results in complex entities being replaced by *even more complex* relational structures. Just as the analysis was *unaffected* by whether the entities implied by the premisses were simples or not, so also the complex structure arrived at by synthesis is unaffected by any later account we might hope to provide of the general nature of its constituents. To put it another way: analysis is primarily analysis of *propositions* as was shown in Chapter Four. Propositions can readily be subjected to the two stages of Russellian analysis even if the exact ontological implications of the current best set of premisses is unclear. All of this is quite in keeping with features of Russellian analysis outlined in earlier chapters.

It has been maintained in this chapter that the synthesis step and its tendency to proliferate relations are major distinctive features of Russellian analysis that are generally overlooked. Features such as these also distinguish Russellian analysis from other methodological principles to which it bears a *prima facie* resemblance. Hypothetico-deductive reasoning[3] and inference to the best explanation (Harman, 1965) are obvious examples here. It is also noteworthy that the features of Russellian analysis outlined in the earlier chapters of this book suggest how the 'paradox of analysis' should be handled. This paradox[4] maintains that all analysis is either false or trivial. If the propositions to be analysed and the analysing propositions are not synonymous, then the analysis is false. If, on the other hand, the two sets of propositions are synonymous, then no information is conveyed by the analysis and it is trivial. Since these are the only two alternatives, we are left with the paradox that all analysis is either false or trivial. As we have found, Russell definitely rejects the notion that the propositions to be analysed and the analysing propositions are synonymous. Instead he offers a detailed account of why, even though the two sets of propositions are not synonymous, the analysis can nevertheless succeed. The synonymity option is, of course, the *foundationalist* view of analysis, which we found Russell decisively rejecting. Russell's *non-foundationalist* alternative leads to premisses or analysing propositions with the features set out in Table 3.1 precisely because no synonymity is involved.

Having detailed the distinctive method that Russell repeatedly applied to his philosophising, it is time to investigate why his philosophical views evolved in the way that they did.

NOTES

1. Pears, 1967, especially Chapters VI, VIII and XI.
2. See also *MPD*: 128.
3. Russellian analysis is equated with hypothetico-deductive reasoning by Maxwell, 1974: 169–82. (See especially p. 181.)
4. See, e.g., Stroll, 1967.

PART TWO

Theories of Space and Time – The Key to Change in Russell's Philosophy

CHAPTER 9

The Impact of Russell's Philosophical Predecessors on His Space and Time Theories

9.1 A Central Preoccupation of Russell's Philosophy

Having identified in Part One an essential *continuity* in Russell's philosophy centred on his method of analysis and its propensity to proliferate relations, we turn in Part Two to an attempt to locate the basis of the major *changes* that take place in this philosophy. The main thesis of Part Two will be that spatial and temporal relations and relata are in many ways the paradigms of relations and relata, and that the major developments in Russell's philosophy stem from the need to change his views on space and time. An evolution in his understanding of space and time becomes an evolution in his understanding of relations and relata in general. Given the importance of relations, highlighted in Part One, such an evolution will clearly have wide ramifications for his overall philosophy. It turns out that Russell's idealist phase, together with the different periods following his revolt against idealism, are each characterised by different views of space and time. These differences constitute, of course, the four distinct phases already outlined in Chapter Four above, each phase representing a significant shift in overall philosophical position. While it is certainly not denied that Russell changed his mind on some points *within* phases, it will be argued that such changes lacked the definitive impact of the new theories of space and time that constitute each distinct phase.

Before examining in detail, in the following chapters, Russell's successive theories of space and time, we need to consider why space and time relations and relata were of such importance as to serve as exemplars of relations for Russell. An obvious reason would be that space and time constitute prime examples of *external relations*,[1] so that their centrality for Russell is only to be expected, given his particular emphasis on external relations. However there is much more to it than this. The major reason for the leading role of space and time relations and relata in Russell's philosophy stems from a constant preoccupation of all of his philosophising, viz. *the need to bridge appearance and reality*. Since this point is vital for what follows, it is important to spell out Russell's conceptions of *appearance* and *reality* and how they differ from those found in the work of some of his immediate predecessors.

It has been common in the history of philosophy for appearance to stand in sharp contrast to reality. In its strongest form, perhaps exemplified in Parmenides, the contrast is the ontological equivalent to illusion or *falsity* as against *truth*. A weaker form of the contrast stems from Plato and is found, somewhat modified, in Kant. In this weaker form, the relation of appearance to reality is akin to that of *opinion* (which may or may not be true) to *truth*. The British Hegelian tradition, on which Russell cut his philosophical teeth, shows a gradual move from the weaker form of the contrast to the stronger one. This tendency culminates in Bradley, whose *Appearance and Reality* is striking for its use of Parmenidean style argument. Accordingly Russell spent much of his idealist apprenticeship in philosophy consigning space, time, motion, matter, change, etc., to the realm of appearance on the grounds that they involved contradictions. As Russell himself testifies:

> In my first rebellion against Hegel, I believed that a thing must exist if Hegel's proof that it cannot is invalid.
>
> (*MPD*: 49).

What this means for the appearance/reality contrast is that from the onset of his realism, Russell accorded to appearance the same *ontological* status as reality, whilst firmly adhering to the *epistemological* primacy of appearance.[2] This, I will argue, provides the distinctive cast to Russell's realism and sets him apart from the mainstream tradition of British empiricism. In identifying why the appearance/reality contrast is so important to Russell's realism, we will at the same time specify the peculiar features of this contrast as it is used by Russell.

It will be helpful for what follows to set Russell's realism against Kant's philosophy. It is a commonplace of the historiography of philosophy that Kant's work represents a reconciliation of insights from both British empiricism and Continental rationalism.[3] As Russell viewed it, Kant was correct to attempt to reconcile the strengths of these two traditions, but went about it the wrong way. Russell's mature philosophy is a series of new attempts to satisfactorily reconcile elements of empiricism and rationalism. As we detail Russell's various realist phases in the following chapters, we will, in effect, be examining the ways in which Russell retraced Kant's footsteps in an endeavour to return philosophy to the correct path from which, as he saw it, Kant had strayed. This strong claim about the basis of Russell's philosophising is, no doubt, controversial. Certainly Russell himself never puts forward the claim as explicitly as one would like. Nevertheless, as this chapter will show, he frequently writes *as though* this is what he saw himself as doing. Indeed, I will argue that a certain series of key principles can definitely be found in Russell's writings and that together they show that his work does bear to Kant the relationship described above. The key principles found in Russell's writings are that:

1. His philosophy deliberately sets out to answer Kant. Specifically, (a) he rejects the undue emphasis on epistemology, which he thinks has bedevilled philosophy since Kant, and (b) he rejects certain doctrines common to both

empiricism and rationalism that were also held by Kant.

2. He derives from empiricism a central doctrine to which he adheres throughout his post-idealist philosophising.
3. He derives from rationalism a central doctrine to which he adheres throughout his post-idealist philosophising.
4. He bases each phase of his post-idealist philosophy on a unique synthesis of the central doctrines derived respectively from empiricism and rationalism (just as Kant's philosophy stemmed from a distinctive reconciliation of the different doctrines that he took over from empiricism and rationalism).

Each of these key principles will now he treated in some detail. In the process, Russell's distinctive use of space and time to bridge appearance and reality will become clear.

9.2 Russell's Philosophy as an Answer to Kant

As previously noted Russell's revolt against idealism was largely prompted by the dawning realisation of its inability to provide a satisfactory account of mathematics. In particular irreducibly relational propositions, proscribed by idealism, appeared to be absolutely fundamental in mathematics. As Russell saw it, the idealist view of mathematics ultimately sprang from the subjectivist doctrines of the *Critique of Pure Reason*, and (as discussed in Section 4.2) he had previously set out to desubjectify Kant, beginning with the *Essay on the Foundations of Geometry*. Thus, although abandoning idealism, Russell continued his earlier efforts to provide a non-subjectivist account of the foundations of mathematics, only this time following the lead of Peano.[4] Hence the logicist programme "to show that all pure mathematics follows from purely logical premisses and uses only concepts definable in logical terms . . . " was viewed by Russell as "an antithesis to the doctrines of Kant . . . " (*MPD*: 57).[5] However if it is well known that Russell's signal contributions in the philosophy of mathematics were intended to displace Kant, it is not so well known that this was *merely part of a wider undertaking*. Russell himself stated, on more than one occasion, that he originally thought of his contribution to *Principia Mathematica* "as a parenthesis in the refutation of Kant" (MMD: 13).[6] Of course *Principia Mathematica* ultimately grew into something much larger than Russell had initially expected. However, the question remains as to whether the wider undertaking to refute Kant was ever carried out, and, if so, how? The answer, I suggest, is that it was and that each of the three phases of Russell's realist philosophy centre on doctrines of space and time that enable appearance and reality to be bridged, thereby providing an alternative to the Kantian approach to philosophy.

Russell's major disagreement with Kant centres on the latter's doctrine that located objects of experience in a *phenomenal* world while relegating things in themselves to the unknowable *noumenal* world. As was discussed in Section 4.1.2, Russell was, at the start of his realism, strongly inspired by a belief in the possibility of attaining absolutely certain impersonal knowledge (a hope that he had abandoned by the 1950's). He was repelled by the Kantian alternatives of either 'knowable but subjective' or 'unknowable but objective'. Of course one has to be careful applying the terms 'subjective' and 'objective' to Kant's philosophy. As Russell himself acknowledged, "Kant's philosophy is difficult and obscure, and philosophers still dispute as to what he meant" (*OP*: 198–9).[7] Accordingly he expected to be accused of misunderstanding Kant. Nevertheless, it is absolutely clear that Russell's basic dispute with Kant lies in his rejection of "the subjectivity of the Transcendental Aesthetic" (MMD: 12) and, in particular, his denial that space, time and relations generally are restricted to the phenomenal world.

(a) Russell's Rejection of the Undue Emphasis on Epistemology Stemming from Kant

One significant outcome of Russell's alternative attempt to reconcile empiricism and rationalism was, as he saw it, to remove epistemology from the position of undue prominence that it had gained in philosophy:

> I reverse the process which has been common in philosophy since Kant. It has been common among philosophers to begin with how we know and proceed afterwards to what we know. I think this a mistake, because knowing how we know is one small department of knowing what we know. I think it a mistake for another reason: it tends to give to knowing a cosmic importance which it by no means deserves, and thus prepares the philosophical student for the belief that mind has some kind of supremacy over the non-mental universe, or even that the non-mental universe is nothing but a nightmare dreamt by mind in its un-philosophical moments.
> (*MPD*: 12).

Here we have a clear repudiation of the various forms of idealism that had flourished in reaction to Kant, in which the phenomena/noumena dichotomy was, of course, replaced by its appearance/reality counterpart – with the difference that reality is objective and knowable while mere appearance is subjective and contradictory. However, Russell objects equally to Kant and the idealists on the grounds that they both share

> . . . a mistaken tendency . . . to allow the description of the world to be influenced unduly by considerations derived from the nature of human knowledge.
> (*HK*: 9).

What is required, according to Russell, is that epistemology be put into proper perspective:

> Ever since Kant, knowledge had been conceived as an interaction, in which the thing known was modified by our knowledge of it, and therefore always had certain characteristics due to our knowledge . . . on the contrary . . . knowledge, as a rule, makes no difference to what is known . . .
> (PTC: 461).

> Cosmically and causally, knowledge is an unimportant feature of the universe; a science which omitted to mention its occurrence might, from an impersonal point of view, suffer only from a very trivial imperfection. In describing the world, subjectivity is a vice. Kant spoke of himself as having effected a 'Copernican revolution', but he would have been more accurate if he had spoken of a 'Ptolemaic counter-revolution', since he put Man back at the centre from which Copernicus had dethroned him.
>
> (*HK*: 9).

Clearly, then, Russell views Kant's attempt to reconcile empiricism and rationalism as enshrining a disastrous emphasis on epistemology, an emphasis originating in Descartes that Russell is determined to discard.[8] Russell also thought that Kant had continued some mistaken doctrines found in both empiricism and rationalism.

(b) *Russell's Rejection of Certain Doctrines Common to Empiricism and Rationalism as well as to Kant*

For a start, Russell is critical of the "subjectivist tendency" that he finds in both traditions, a tendency that was to inspire Hegel's reaction (*HWP*: 75–7). However, Russell's main target here is the doctrine that all genuine propositions are of the subject-predicate form, a doctrine that leads inevitably to the further doctrine that relations are purely subjective. As Russell saw it, both doctrines are prominent in empiricism and rationalism,[9] are basic to Kant's system and underlie much of the idealist tradition inspired by Kant. Leibniz is largely blamed here since, as we will find shortly, in Russell's view, he refrained from accepting the logical outcomes of his own theories. Leibniz's failure to countenance the objective space and time allegedly implied by his doctrines is attributed by Russell to his dogmatic adherence to the belief that all propositions are of the subject-predicate form, with the consequent denial of objectivity to relations generally. According to Russell, Kant's reconciliation of empiricism and rationalism follows Leibniz in its commitment to the primacy of the subject-predicate form whilst "boldly" facing "the consequences . . . of the subjectivity of space" (*PL*: 99). The major consequence is, of course, the unknowable thing in itself (*PL*: 15). Kant's phenomenal and noumenal worlds are clearly descendants of Leibniz's realms of nature and grace. This commitment to the subject-predicate form and the consequent downgrading of relations was then carried over into the various versions of absolute idealism spawned in reaction to Kant. The rejection of these doctrines forms part of Russell's new beginning that was to be a viable alternative to Kant.

What did Russell take over from the two traditions? From rationalism Russell could be said to support an emphasis on mathematics and logic with the innovation that relational propositions are viewed as fundamental. This is balanced by a piecemeal approach to philosophical problems that is more characteristic of empiricism.[10] Russell also rejects the category of substance, which typifies rationalism. This follows his finding, in *A Critical Exposition of the Philosophy of Leibniz*, that it had become "necessary . . . to base metaphysics on some notion other than that of substance – a task not yet accomplished" (*PL*: 126).[11] Russell's contribution was to base metaphysics on relations rather than on substance. However, the major influence of the two traditions on Russell can be seen in the distinctive doctrines that he derived from each and retained throughout his subsequent career. Russell usually credits Berkeley and Leibniz[12] respectively for inspiring these doctrines and to these we now turn.

9.3 Russell's Central Doctrine Derived from Empiricism

Berkeley convinced Russell that *all qualities are secondary*. As Russell put it in 1946:

> . . . Berkeley advances valid arguments in favour of a certain important conclusion, though not quite in favour of the conclusion that he thinks he is proving. He thinks he is proving that all reality is mental; what he is proving is that we perceive qualities, not things, and that qualities are relative to the percipient.
>
> (*HWP*: 624).[13]

All qualities are relative to the percipient in that they are sense-data and therefore secondary. Alleged primary qualities like shape are not exempt here:

> The 'real' shape is a shape in physical space, which has no more resemblance to visual space than light-waves have to colour. Shape as the sense-datum is in just the same position as colour.
>
> (NSD: 187).

This denial of the primary/secondary distinction is not a claim that 'all qualities are in the mind' precisely because, as was emphasised at the start of this chapter, appearances (sense-data) have the same ontological status as reality in Russell's view. One implication of this is that sense-data (percepts in the Modified Empiricist phase) *can* tell us something of their causal antecedents. As Russell points out

> . . . spatial order reduces to relations, and these relations have certain logical properties in virtue of which they generate a three-dimensional manifold. I suggest that the apparent shape 'corresponds' as a rule to a real shape, due to relations having similar *logical* properties. But it is a case of *correspondence*, not identity, just as in the case of colours and their physical correlates.
>
> (NSD: 187).

This feature of the logical properties of relations provides Russell with a general principle which can be seen as his substitute for the misconceived primary/secondary quality distinction. The principle is that we

> . . . can know the properties of the relations required to preserve the correspondence with sense-data, but we cannot know the nature of the terms

> between which the relations hold. . . . Thus . . . although the *relations* of physical objects have all sorts of knowable properties, derived from their correspondence with the relations of sense-data, the physical objects themselves remain unknown in their intrinsic nature. . . .
>
> (*PP*: 32, 34).[14]

This general principle was upheld by Russell throughout the various phases of his realism[15] and was, of course, the basis of the thesis of the *epistemological primacy of relations*, which was discussed in Chapter Six.

Of course, this general principle immediately places Russell at variance with Kant. As against the Kantian position that the raw material of sensation is the only aspect of experience due in part to the external world, with relations, whether spatial, temporal, or whatever, being supplied by the mind, Russell adopts a contrary view. According to him qualities are subjective, although *externally caused*, whilst the relations between their instances closely represent the structure of the external world.

Kant's weakness, says Russell, is that he allows:

> . . . that the mind orders the raw material of sensation, but never thinks it necessary to say why it orders it as it does and not otherwise.
>
> (*HWP*: 687).

The answer, according to Russell, is that the order in the mind is a more or less exact representation of the order in the external causes of the sensations. As we will find, a lot hinges on the 'more or less' here, with the representation becoming less exact as Russell's realism proceeds through its various phases.

Our examination of Russell's central doctrine derived from empiricism has shown it to be of some help in the project of bridging appearance and reality. From the nature of appearance, the principle enables us to infer something of the nature of its causal antecedents, in particular their spatial and temporal relations. To see how the bridging of appearance and reality can be taken further, we need to turn to the views inspired by Leibniz.

9.4 Russell's Central Doctrine Derived from Rationalism

The doctrine that Russell derives from Leibniz stems from a modification of the theory of monads. Leibniz, as Russell understood him, held that spatial relations hold "only between simultaneous objects of perception of each monad", but "do not hold between monads", thereby ensuring that "space is properly subjective" (*PL*: 122). But, as Russell argued at length in his major study of Leibniz's philosophy, Leibniz failed to follow his own arguments where they led, viz., to an objective space occupied by the monads:

> . . . the perceptions of different monads differ, owing to the difference of the points of view; but points of view are mathematical points, and the assemblage of possible points of view is the assemblage of possible posi-

> tions. Thus Leibniz had two theories of space, the first subjective and Kantian, the second giving an objective counterpart, *i.e.* the various points of view of the monads. The difficulty is, that the objective counterpart cannot consist *merely* in the difference of points of view, unless the subjective space is *purely* subjective; but if it *be* purely subjective, the ground for different points of view has disappeared, since there is no reason to believe that phenomena are *bene fundata.*
>
> (*PL*: 122).

Similar considerations apply to Russell's analysis of Leibniz's theory of time. Hence Russell's conclusion

> . . . that Leibniz, more or less unconsciously, had two theories of space and time, the one subjective, giving merely relations among the perceptions of each monad, the other objective, giving to the relations among perceptions that counterpart, in the *objects* of perception, which is one and the same for all monads and even for all possible worlds.
>
> (*PL*: 129).

As noted earlier, Russell's diagnosis of Leibniz's failure to reach these conclusions is based on the latter's dogmatic rejection of non-subjective relations, a rejection in turn grounded in a fundamental commitment to the primacy of the subject-predicate form for propositions.

So, throughout his realism, Russell holds that besides subjective spaces and times that are private to percipients, there is objective (or physical) space and time.[16] If the doctrine derived from empiricism assists the project of bridging appearance and reality by enabling us to infer some of the features of the causal antecedents of sense-data, what contribution does this doctrine inspired by Leibniz make to the said project? Fairly obviously, by correlating the contents of private worlds of appearance with the contents of physical space and time, one would have gone a long way towards bridging appearance and reality and providing an alternative to Kant. *How* this is done varies markedly between the three phases, so I will leave the details for the following chapters. However, the doctrine inspired by Leibniz serves another important function in Russell's philosophy which is worth pursuing here. I refer to his desire to reconcile the worlds of appearance and of science.

It was pointed out earlier in this chapter that Russell was critical of the undue emphasis on epistemology that was Kant's legacy to subsequent philosophy. In his view, "knowing how we know is one small department of knowing what we know."[17] Not surprisingly, Russell takes science to be the best example of knowledge about the world: "In ontology I start by accepting the truth of physics" (RTC: 700).[18]

To the critic who objects that science worship is out of place in philosophy, especially since we cannot tell in advance which parts of present day science will be rejected ultimately, Russell replies that no more secure starting point is available:

> . . . I have no doubt that, although progressive changes are to be expected in physics, the present doctrines are likely to be nearer to the truth than any rival doctrines now before the world. Science is at no moment quite right, but it is seldom quite wrong . . . It is, therefore, rational to accept it hypothetically. . . .
>
> (*MPD*: 13).

Yet despite these resolutions to take for granted the better examples of knowledge, Russell is still reputed in many quarters to be a prime case of an epistemological sceptic. Looking back over his career in 1943, Russell himself lamented that:

> In some respects, my published work, outside mathematical logic, does not at all completely represent my beliefs or my general outlook. Theory of knowledge, with which I have been largely concerned, has a certain essential subjectivity; it asks 'how do *I* know what I know?' and starts inevitably from personal experience. Its data are egocentric, and so are the earlier stages of its argumentation. I have not, so far, got beyond the earlier stages, and have therefore seemed more subjective in outlook than in fact I am. I am not a solipsist, nor an idealist; I believe (though without good grounds) in the world of physics as well as in the world of psychology.
>
> (MMD: 16).[19]

What, at first sight, might appear to be a simple case of a mismatch between intention and outcome, takes on a different meaning when viewed as an ongoing attempt to bridge appearance (the world of private perceptions) and reality (the world of science). Russell's problem was, as the last quotation suggests, that the bridging turned out to be more difficult than expected. That there are three post-idealist phases of Russell's philosophy, outlined in the following chapters, is largely due to the fact that the reality that he thought science postulated kept changing. Specifically, Russell's understanding of physical space and time changed, so that subjective space and time are correlated with different entities in each of the different phases. However, throughout all of this, the doctrine of the dual spaces and times derived from Leibniz was held unwaveringly.

9.5 Synthesis of the Empiricist and Rationalist Doctrines

As will be demonstrated in subsequent chapters, each phase of Russell's post-idealist philosophy is based on a synthesis of the central doctrines derived respectively from empiricism and rationalism. Fundamental to such syntheses are space, time, matter and force, which Russell took to be the four fundamental concepts enshrined in Newtonian physics.[20] Hence these four concepts, or, rather, their successors in post-Newtonian physics, will be crucial to philosophy. Russell's initial idea was that the doctrine of dual spaces and times, inspired by Leibniz, would handle the first three concepts, since the contents of the respective spaces and times would be amenable to simple correlation. Likewise the doctrine inspired by the empiricist tradition would handle matter and force (or cause), since the relational properties of sense-data would mirror the relational properties of their material causes. Russell's comment in another context is appropriate here: "My views at that time had a kind of morning innocence which they lost in the labour and heat of the day" (*MPD*: 118). The results of the labour and heat of the day are detailed in the following chapters. However Table 9.1 does, I think, reveal some of the complexity that Russell encountered. It shows the changing characteristics of physical space and its correlate, the space of appearance, as Russell's philosophy developed.

Table 9.1. Russell's changing post-idealist views on the two spaces

	Spatial Appearance	Spatial Reality (Physical space)
Platonist phase	Relative and subjective	Absolute and objective
Empiricist phase	Relative and subjective	Relative and objective
Modified empiricist phase	Absolute and subjective	Space-time instead of space (and time)

In the latter part of this chapter, I have stressed Russell's determination to develop an ontology that incorporates both sense-data (percepts) and the entities of physics, whilst at the same time ensuring that knowledge based on the world of appearance is consistent with scientific knowledge. This has, I trust, supported and clarified my claim, early in this chapter, that the distinctive cast of Russell's realism is due to his according appearance the same *ontological* status as reality, whilst firmly adhering to the *epistemological* primacy of appearance. I have also strongly urged in this chapter that Russell's philosophising can be viewed fruitfully as a rebellion against Kant. I take comfort in this view from Russell's own rejoinder to Alan Wood's suggestion that one might sum up Russell's "public career as a philosopher, briefly and crudely, as: From Kant to Kant":

> I cannot subscribe to this formula. My final views are less Kantian than Alan Wood supposes. I will mention two points. First: though the external world is probably not quite like the world of perception, it is connected with the world of perception by correlations, which are impossible in a philosophy which regards time and space as subjective. Second: the principles of non-deductive inference which I advocate are not put forward as certain or *a priori*, but as scientific hypotheses.[21]

Notes

1. Space and time as exemplars of external relations is a notion with a long history. Hume's two-fold classification of the seven sorts of relations into "relations of ideas" and "matters of fact" (*Treatise*, Book I, Part III, Section I) is a likely ancestor of the internal/external relations dichotomy. For Hume, relations of space and time, along with identity and causation,

are external, i.e. "matters of fact". (For discussion of Hume's distinction as an early version of the internal/external distinction see Armstrong, 1978, Vol. 2: 84ff.)

2. NSD: 187–8 is very emphatic on this point about appearance and reality.
3. See, e.g., Copleston, 1964: 218; Scruton, 1982: 14; Wilkerson, 1976, Ch. 1.
4. For discussion of Peano's influence on Russell see Grattan-Guinness, 1976 and Rodriguez-Consuegra, 1991.
5. Cf. *MPD*: 56, where Russell claims that Peano's work on geometry showed "the needlessness of Kant's *Anschauung*. . . . "
6. See also *MPD*: 57. Russell repeatedly stated that philosophy needed to start again with a rejection of Kant. See, e.g. MMD: 12–3 and *MPD*: 12.
7. See also *PP*: 85.
8. Rorty (1980: 8) makes a similar plea for modern philosophy to radically reorient itself away from the dominance of the Cartesian-Kantian epistemological tradition. His major criticism of this tradition is that it has philosophy "judging other areas of culture on the basis of its special knowledge of the 'foundations' of these areas". Rorty includes Russell in the epistemological tradition. Ironically, Rorty's remarks are very similar to what Russell was saying fifty years earlier. (See, e.g. PTC: 460 and *HWP*: 783–9.) Yet it is not surprising that Rorty is obviously unaware of Russell's position since his book makes it clear that he adopts a *foundationalist* interpretation of Russell, an interpretation that was shown to be mistaken in Chapter Four above.
9. *Papers* 7: 93: "The polemic of the English empiricists against 'substance' as the unknown substratum of the predicates really agrees with its opponents as to the logical nature of subject-predicate propositions."
10. Russell emphasises both of these influences in his characterisation of his own philosophy in the final chapter of *History of Western Philosophy* (*HWP*: 788).
11. See also *PL*: 5ff.
12. Interestingly, in Chapter Two of *MPD*, entitled "My Present View of the World", Russell describes his position as a reaction to Kant and specifically invokes Leibniz (pp. 17–8) and Berkeley (p. 20) as the major influences on his position. (See also *Auto.*: 190, 395.)
13. See also *PP*: 38–41; *AMa*: 132 and *MPD*: 20.
14. See also *OP*: 103–4; *AMa*: 227 and Ch. XXIV; *MPD*: 13, 19 and *HWP*: 688.
15. Strictly speaking 'physical objects' (Platonist phase) would be replaced by 'sensibilia' (Empiricist phase) or 'events' (Modified Empiricist phase). Likewise 'sense-data' (earlier phases) become 'percepts' (Modified Empiricist phase). But the general principle remains.
16. Strictly speaking, there is a plurality of private spaces: visual, tactile, olfactory, etc. (*PP*: 29ff.). Like most philosophers, Russell concentrates on the visual. Physical space and time later become space-time.
17. E. Gellner agrees with Russell's critique of the emphasis in modern philosophy on epistemology. See Magee, 1982: 253.
18. See also LA: 175. Russell's faith in science was noted in Chapter Four.
19. Russell adds that a future work (what became *HK*) would seek to remedy this situation.
20. See, e.g., *MPD*: 12–3 and UCM: 81.
21. Wood's claim and Russell's rejoinder are in *MPD*: 194. Russell, of course, expressed strong support for the remainder of the Wood material.

CHAPTER 10

Space and Time in the Platonist Phase (1899–1913)

Russell's views on space and time in the Platonist phase were, of course, preceded by detailed work on this topic in the neo-Hegelian phase. As we found in Chapter Four, Russell analysed geometry, physics and pure mathematics and the dialectical relationships between them. He repeatedly identified paradoxes and antinomies which he viewed as inevitable contradictions generated by the limited and partial assumptions on which each special science rested.[1] His hope was that these contradictions might be resolved by a dialectical transition through the sciences, e.g. the contradictions in geometry might be resolved in physics. However, the final elimination of contradiction and paradox would require a dialectical transition or supercession to a wider, over-arching science whose special subject matter would be the Absolute. Thus, for the neo-Hegelian Russell, physical space and time have no ultimate reality. At best, reinterpretation of our understandings of space and time might tell us something about the Absolute. As *Papers* 2 testifies, Russell continued to work on space and time right till the end of the neo-Hegelian phase, hence it comes as no surprise that real, perduring points and instants were prominent abstract entities in his Platonist ontology.

Before embarking on a detailed consideration of space and time in Russell's Platonist phase, it is important first to recall some central features of his work in this phase. As with all versions of Platonism, Russell's one postulates a realm of abstract entities outside of space and time, the world of *being* or *subsistence*. He describes this "world of being" as "unchangeable, rigid, exact, delightful to the mathematician, the logician, the builder of metaphysical systems, and all who love perfection more than life" (*PP*: 100). In *Principles of Mathematics* Russell had populated the world of being with a diverse range of entities. As we found in Chapter Seven, denoting concepts, many of which themselves failed to denote, accounted for much of this world of being. Then developments such as the theory of descriptions led to a gradual pruning of this range of entities assigned to the world of being. Chief amongst the entities eliminated were denoting concepts, propositions and classes. Towards the end of the Platonist phase in *Problems of Philosophy* universals had become the prime occupants of the world of being (*PP*: 100).[2] However these changes in the *range* of entities are incidental to the main point that throughout his Platonist phase (1899–1913), Russell was clearly committed to the world of being.

In contrast to the world of being, the world of *existence* was "fleeting, vague, without sharp boundaries, without any clear plan or arrangement" and contained "all thoughts and feelings, all the data of sense, and all physical objects . . . " (*PP*: 100). According to Russell, both worlds "have the same claim on our impartial attention, both are real, and both are important to the metaphysician" (*PP*: 100).[3] For our present concerns, this raises two important questions. Firstly, and obviously, what are the relations between the world of being and the world of existence? Secondly, given that both worlds together constitute reality and given also that appearance (in the sense specified in the previous chapter) is a sub-class of reality, how is appearance related to reality? Since appearances (sense-data) are only *part* of the world of existence, the two questions are obviously distinct. It turns out, however, that the answers to both questions hinge on Russell's doctrines of space and time. In the former case, the nature of the relations between the two worlds has been a recurring difficulty for the various versions of Platonism, beginning with Plato's own wrestling with the metaphors of imitation and participation. Russell's solution to the difficulty depends on space and time being in the unique position of 'having a leg' in each world, this despite the world of being not being in space and time. Once this machinery for answering the first question is set up, Russell's answer to the other question, concerning the relations between appearance and reality, follows fairly straightforwardly. So, then, the immediate task is to outline Russell's Platonist theories of space and time with particular attention to the way that they are used to link the worlds of being and existence.

10.1 Relations Between Being and Existence

In *Principles of Mathematics*, the major early work of the Platonist phase, Russell was primarily concerned with the *logical analysis* of pure mathematics.[4] Geometry as a branch of pure mathematics is independent of the study of physical space – it is "*the study of series of two or more dimensions*" and is "a pure *a priori* science" which is ". . . indifferent . . . to the question whether there exist (in the strict sense) such entities as its premisses define" (*PoM*: 372). Clearly the world of being is the domain of this logical analysis of pure mathematics. However, Russell's intellectual curiosity is such that in *Principles of Mathematics* he cannot help himself from straying to matters "irrelevant to pure mathematics" (*PoM*: 445–6), such as the nature of physical space, thereby shifting his attention to the world of existence. Russell's conclusion about physical space is that the arguments against the absolute theory, with its commitment to the existence of *points*, are "one and all fallacious". So while pure geometry deals with the multiplicity of series in the world of being having more than one dimension, in the world of existence, "at least one series of several dimensions exists, namely, space" (*PoM*: 461). Similar considerations apply to time. Since the central doctrines of *Principles of Mathematics* represent a total repudiation of the Kantian view that arithmetic depends on time (*PoM*: 458), it is not surprising to find time receiving less consideration than space in that work. Nevertheless, since Russell is concerned with, e.g.,

> . . . the analysis of rational Dynamics considered as a branch of pure mathematics, which introduces its subject-matter by definition, not by observation of the actual world. . . .
>
> (*PoM*: 465).

enough is said to give us Russell's theory of time in broad outline. Just as space in the world of being takes the form of series of two or more dimensions, so its equivalent for time is a certain one-dimensional series (*PoM*: 473). Likewise, Russell's remarks say enough about the world of existence to make it clear that he favours an absolute theory of time with its commitment to the existence of *instants* (*PoM*: 265).

Fortunately the sketchy treatment of physical space and time in *Principles of Mathematics* is supplemented by two important 1901 papers: "On the Notion of Order" and "Is Position in Time and Space Absolute or Relative?"[5] These sources between them reinforce the theories of space and time that underlie *Problems of Philosophy*, Russell's major treatment in the Platonist phase of the worlds of being and existence and the relations between them. It has already been intimated that Russell's solution to the problem of the nature of the relations between the two worlds depends on space and time uniquely 'having a leg' in each. We are now in a position to give substance to this metaphor.

Although "both being and existence . . . belong to empty space" (*PoM*: 449),[6] and also to the instants that constitute time (PTS: 261–3), according to Russell, "being alone is relevant" to settling whether space and time are absolute or relative (*PoM*: 449).[7] Of course, for Russell in 1903, the denoting concepts for points and instants would have being even if the relative theories were true. However Russell's view is that logical analysis of pure geometry and of temporal series will decide whether points and instants are essential to the being of space and time. His conclusion that they are entails that, during the Platonist phase, he is committed to absolute theories of space and time.[8] However, in our present task of showing how Russell's space and time theories supply the connections between the worlds of being and existence, it is not the *being* of the points and instants that is vital (as we will find, it is, rather, the being of the special serial relations that they instantiate which matters). Instead, it is the *existence* of points and instants that is of central interest for connecting the two worlds. Whereas *being* was pertinent to the refutation of relative theories of space and time, *existence* underlies Russell's rejection of claims that physical space and time are subjective (*PoM*: 449).[9] Hence in the Platonist phase, Russell is an objective realist about points and instants.

Not only do points and instants exist, they enjoy a special pre-eminence amongst existents. According to Russell, existents comprise

> . . . four great classes: (1) instants, (2) points, (3) terms which occupy instants but not points, (4) terms which occupy both points and instants.
>
> (*PoM*: 465).[10]

From this it is clear that points and instants are the most basic of existents since all other existents stand in a relation of occupation to them. For (3), Russell had in mind "mental qualities" which ". . . are in time, but are often held to be not in space" (PTS: 264). An example would be *percepts* – by contrast, *concepts* would be in neither space nor time, and hence would not be existents (RUP: 167–8).[11] Points and instants are not only basic existents in that they are occupied by all other existents, they are also basic from the point of view of analysis:

> What is meant by *occupying* a point or an instant, analysis cannot explain; this is a fundamental relation, expressed by *in* or *at*, asymmetrical and intransitive, indefinable and simple.
>
> (*PoM*: 465).

The prime occupants of points and instants are "bits of matter" and instances of qualities (*PoM*: 465–6) (whiteness is a universal having being but not existence, and therefore not in space or time, while instances of whiteness are particulars occupying points and instants). Such occupation of a place at a time constitutes "an irreducible triangular relation" (*PoM*: 472).

All of this talk about "irreducible relations" which "analysis cannot explain" apparently clashes with one of the major findings of Part One that Russellian analysis is never final. However, such talk of irreducibility should be interpreted as merely provisional claims, subject to revision in accordance with the general principles of Russellian analysis. After all, Russell offers no argument to show that occupancy must be an irreducible relation. In the heat of analysis, it is not unexpected that its provisionality should be left unexpressed. I would also stress that the "irreducible relations" that Russell is discussing occur in the *world of existence*. As the preface and early sections of *Principles of Mathematics* make abundantly clear, Russell's general logical analysis of mathematics concerns the *world of being* and closely accords with the non-finality thesis of Part One. I suggest that the occupation of points and instants being the limit of analysis in the world of existence adds further support to my claim that space and time form the link between the worlds of being and existence. Russell, true Platonist, will explain the world of existence in terms of the world of being. Thus philosophical analysis of results concerning the world of existence will only go so far without drawing on considerations about the world of being.

Whilst occupation of points and instants is a limit for the world of existence, we find that the serial relations instantiated by the points and instants have *being* but not *existence*. So, whereas points and instants as basic existents gave space and time a firm grounding in the world of existence, it is the *relations* between the points and instants that provide the connection with the world of being. Likewise, just as points and instants are pre-eminent amongst existents for the reasons already noted, the serial relations instantiated by the points and instants enjoy "a distinguished place among entities" in the world of being (ONO: 309). This is because these serial relations are prime examples of what Russell calls *independent series*. The order of most sets of terms is more or less arbitrary. Since they have no intrinsic order, we can highlight any order we like. But

> . . . this possibility depends entirely upon the fact that there are sets of terms having intrinsic order, with which any other set of terms can be correlated. A casual collection of terms may be ordered by counting, in which case they are correlated with the integers; by speech, in which case they are correlated with a series of times; or by writing, in which case they are correlated with a series of places. But the order arises, in each case, from the intrinsic order of the integers, the times, or the places respectively. These have an order independent of our caprice – they form what I shall call *independent* or *self-sufficient* series. The casual terms correlated with them form, on the contrary, only a *series by correlation*. . . . Thus all orders by correlation are logically dependent upon intrinsic orders.
>
> (ONO: 291–2).[12]

As a consequence of this, "geometrical relations hold primarily between points, and only by correlation between particles. Thus geometrical propositions become timeless . . . " (PTS: 282). Unlike other existents, points and instants are changeless terms instantiating unchanging, independent serial relations. This provides another respect in which points and instants are unique amongst existents: they appear to be the *only changeless existents* that are terms of independent serial relations. Other terms of such relations either lack existence (such as the integers) or have existence but are not changeless (such as series of colour instances or of sounds). In establishing series by correlation for existents other than points and instants, the vital role of the "irreducible triangular relation" of *in* or *at*, that we found existents bear to points and instants, is obvious.

What, then, of the central feature of the independent serial relations instantiated by points and instants, viz. that they have *being* but not *existence*? This stems from an important asymmetry between Russell's views on properties and relations dating from early in the Platonist phase.[13] Properties, being universals, have being but not existence. However, *instances* of properties are existents, i.e. they are located in space and time. Thus instances *of* whiteness can be in many places at the same time. Properties, in short, can be *particularised*. Likewise, relations are universals and therefore have being but not existence. However, unlike properties, relations don't have instances, i.e. relations are *not* particularised.[14] So while whiteness can be found at each of the places and times occupied by its particularisations, there "is no place or time where we can find the relation 'north of' " that relates (say) Edinburgh to London (*PP*: 98). While the *terms* of relations can have existence, the relations themselves belong purely to the world of being. Here is Russell's answer to Plato's problem of the connection between the two worlds. Space and time are the exemplars of this connection because they comprise unique existents instantiating prime examples of independent serial relations. Thus, in a very special way, space and time have a 'leg in each world'.

In the discussion of logical constructions in Part One, the mistake of taking them to be mere classes of particulars was emphasised.[15] We found that in many important cases the relations were an equally vital part of the logical construction. We also found that Russell himself had unwittingly contributed to the widespread confusion on this issue. In the same way, *points* (and *instants*), as well as the *relations* between them, would seem to be equally vital constituents of space and time in the Platonist phase. Yet we find Russell declaring that, e.g., space ". . . is, in fact, essentially a class [of points] . . . " (*PoM*: 443). Fortunately Russell is not here repudiating the views presented in this chapter. In the same paragraph he elaborates as follows:

> Relations are, of course, quite essential to a space . . . and if relations are essential to the definition of a space, there must be some class of relations having a field which is the space.

So the points, as the field of the special independent serial relations, constitute the space. Russell's emphasis on space as "essentially a class of points" has a polemical intent.[16] Firstly, Russell in *Principles of Mathematics* is attacking relational accounts of space derived from Leibniz and his followers, hence it is the points rather than the relations that need emphasising. Secondly, Russell is striving to avoid misunderstandings arising from the rather complicated theory of wholes and parts that underlies much of *Principles of Mathematics*. According to this theory, there are two distinct sorts of wholes, which Russell calls *unities* and *aggregates*. The main feature of unities is that they "are not completely specified when their parts are all known", whereas aggregates are completely specified when all their simple parts are known (*PoM*: 140). Russell uses as an illustration of the distinction the proposition "*A* differs from *B*" where *A* and *B* are simple terms.

> The simple parts of this whole are *A* and *B* and difference; but the enumeration of these three does not specify the whole, since there are two other wholes composed of the same parts, namely the aggregate formed of *A* and *B* and difference, and the proposition "*B* differs from *A*". In the former case, although the whole was different from all its parts, yet it was completely specified by specifying its parts; but in the present case, not only is the whole different, but it is not even specified by specifying its parts.
> (*PoM*: 140).

For a number of reasons[17] Russell was concerned to emphasise that space and time are aggregates, not unities. This, then, is the context in which space is described as "essentially a class of points", albeit a class that is the field of a special independent serial relation. As we have already found, from the viewpoint of *existence*, space *is* a class of points since the relations belong to the world of being. The same considerations also apply, of course, to time.

One further feature of points (or instants) and their relations is worth stressing. It is well known that Russell frequently held up the doctrine of *external relations* as the focus of his revolt against idealism.[18] I suggest that the relations governing points (or instants) are paradigmatic examples of *external* relations.

This is obvious from a consideration of what the doctrine of external relations entails:

> . . . there are such facts as that *x* has the relation *R* to *y*, and that such facts are not in general reducible to, or inferable from, a fact about *x* only and a fact about *y* only: they do not imply that *x* and *y* have any complexity, or any intrinsic property distinguishing them from a *z* and a *w* which do not have the relation *R*.
>
> (ERB: 355).[19]

Since points (or instants) lack complexity and do not differ in respect of intrinsic properties, the relation between any pair of points (or instants) is evidently an external relation *par excellence*. The same considerations clearly apply to the irreducible relations of *in* or *at* that we have found existents bear to points and instants.

So far in this chapter it has been argued that Russell used space and time to relate the worlds of being and existence. The unique features of points and instants are vital to this project. The following table highlights how points and instants, though existents, share some of the features of being that mark it off from all other existents. Thus points and instants are, in a sense, on the border of the two worlds.

Table 10.1. Points and instants in relation to being and existence

World of being	World of existence	
	Points and instants	Existents other than points and instants
Not in space and time Changeless Domain of independent serial relations	Not in space and time Changeless Fall under independent serial relations	In space and time Changing Fall under dependent serial relations, i.e. series by correlation (typically)

As shown in Table 10.1, points and instants themselves are not in space and time, since to be in space and time is to have an "irreducible triangular relation" of *at* or *in* to a point and an instant. Whereas points and instants, in common with the world of being, are changeless, existents other than points and instants are subject to change, both in terms of their properties and in terms of their relations to points and instants.

10.2 Bridging Appearance and Reality

Having seen how Russell uses space and time to link the worlds of being and existence, we are now in a position to carry out the more important task of showing how space and time are used to bridge *appearance* and *reality* in the

Platonist phase. We saw briefly how this was to be done in the previous chapter. Though "... the physical world and the sensible world have no point in common, ... there is ... a more or less exact correspondence between the two" (AR: 136). That is, the space and time of appearance are such that their contents closely correlate with the contents of physical space and time. In particular, certain *relations* of the contents of the space and time of appearance will correspond to relations amongst their causal antecedents in physical space and time, even though we know nothing of the *properties* of those causal antecedents. We can now draw on our findings in this chapter on being and existence to elucidate the details of appearance/reality bridging in the Platonist phase. In the following I use space, more particularly visual space, as the example since this is the case Russell discusses in most detail. The application of the same principles to time is, I think, clear enough.[20]

As suggested previously in Table 9.1, in the Platonist phase Russell holds physical space to be *absolute* and *objective*, whilst the space of appearance is *relative* and *subjective*. Thus the two spaces to be correlated possess rather disparate features. This presents *prima facie* difficulties for the correlation because

> ... the space of perception is not infinitely divided, and does not consist of points, but is composed of a finite though constantly varying number of surfaces or volumes, continually breaking up or joining together according to the fluctuations of attention. If there is a 'real' geometrical space corresponding to the space of perception, an infinite number of points in the geometrical space will have to correspond to a single simple entity in the perceived space.
>
> (RUP: 174).

However, before considering how Russell resolves this problem, the claim that physical space is viewed as absolute throughout the Platonist phase needs to be defended since some doubt on this point is evident in the literature. Nusenoff, in a discussion of *Problems of Philosophy*, attributes to Russell the view that physical space is "relational rather than absolute" (1978: 66). Likewise Eames sees Russell's "attempt to analyse the data of sensation in such a way that they could provide a basis for the 'construction' of space, time, matter and cause ... " as his main preoccupation in technical philosophy "between 1910 and 1913" (1984: xx). According to Eames, *Problems of Philosophy* is "a first sketch" of a theory of knowledge in which points, moments and matter are analysed in terms of sense-data (1984: xxi). If Nusenoff and Eames are right, then I ought to be placing *Problems of Philosophy* in the Empiricist phase rather than the Platonist one. In Nusenoff's case no evidence is given to support his claim, though it is evident that he is anxious to maximise the continuity between *Problems of Philosophy* and what he takes to be Russell's phenomenalist works of 1914–15. The basis for Eames' claim seems to be her view that during "the winter and spring of 1912 Russell wrote a good deal" on the topics that "bore fruit in his 1914 publications" (1984: xx). Now Eames may well be right on this last point. However, I want to urge that there are compelling reasons for doubting that

Russell had the notion of constructing points and instants out of sense-data when he wrote *Problems of Philosophy* in 1911.

For a start there is the textual evidence. The discussion of space and time in *Problems of Philosophy* contains no hint of constructing points and instants. Instead everything that Russell says is perfectly consistent with his supporting absolute theories of physical space and time. (The same applies to "On the Relations of Universals and Particulars" which was read as his Presidential Address to the Aristotelian Society at the time *Problems of Philosophy* had just been completed.) So, for example:

> We can know . . . that the earth and moon and sun are in one straight line during an eclipse, though we cannot know what a physical straight line is in itself, as we know the look of a straight line in our visual space.
>
> (*PP*: 31).[21]

The Russell here talking about the nature of "a physical straight line in itself" obviously thinks of the line as a compact one dimensional series of unobservable points, not as a logical construction out of sense-data. This conclusion is strongly supported by the Whitehead scholar Victor Lowe. In an article on Whitehead's August 1911 critical comments on the final typescript of *Problems of Philosophy*, Lowe concludes as follows on the application of the method of construction to points and instants:

> Whitehead's criticism of *The Problems* does not mention the principle of substituting constructions for inferences, and I doubt that he had yet shown this use of it to Russell.
>
> (Lowe, 1974: 9).[22]

There is a further reason for thinking that Russell was unaware of the notion of constructing points and instants out of sense-data when he wrote *Problems of Philosophy*. Simply, since Russell repeatedly informs us of the vital significance of Whitehead's discovery and its effects on his own philosophical development, it is hard to imagine him not proclaiming it in *Problems of Philosophy* had he known of it. So, for example:

> As regards points, instants, and particles, I was awakened from my 'dogmatic slumbers' by Whitehead. Whitehead invented a method of constructing points, instants, and particles as sets of events, each of finite extent. This made it possible to use Occam's razor in physics in the same sort of way in which we had used it in arithmetic. I was delighted with this fresh application of the methods of mathematical logic. It seemed to suggest that all the smoothness of the concepts used in theoretical physics could be attributed to the ingenuity of mathematicians rather than to the nature of the world.
>
> (*MPD*: 77).[23]

In the Preface to *Our Knowledge of the External World* we are told that:

> . . . I have been made aware of the importance of this problem by my friend

and collaborator Dr Whitehead, to whom are due almost all the differences between the views advocated here and those suggested in *The Problems of Philosophy*. I owe to him the definition of points, the suggestion for the treatment of instants and 'things', and the whole conception of the world of physics as a *construction* rather than an *inference*.

(*OKEW2*: 8).

The differences between the two works that Russell here stresses are, of course, essentially the differences between what I have called the Platonist and Empiricist phases. So, adhering to the principle set out in Chapter One of this book, according to which phases in Russell's work are dated by significant published works, the Platonist phase includes the 1912 *Problems of Philosophy*. This leaves the Empiricist phase to start with the 1914 works in which points and instants are logical constructions out of sense-data.

So, then, physical space in the Platonist phase is *absolute* and *objective*, i.e. it consists of an infinite number of existent points in which ". . . any matter occupying more than a point of space is regarded as at least theoretically divisible" (RUP: 174). The question is how is this to be correlated with the space of appearance that is *relative* and *subjective*, i.e. it does not consist of points? Rather ". . . perceived spatial relations . . . hold between the parts of a single complex object of perception . . . Perceived spatial relations, therefore, must have a certain roughness, not the neat smooth properties of geometrical relations between points" (RUP: 175).

Take the simple case of a small number of physical objects in a straight line. As we found earlier in this chapter, the points of Newtonian space which Russell believes physics requires, are existents falling under a particular series of three dimensions. The series has, of course, the appropriate kinds of compactness and continuity and is an *independent series*. We have just seen that each physical object is regarded by Russell "as at least theoretically divisible" so that each particle of its matter has an unanalysable *at* relation to a point. In this way the group of physical objects forms a *series by correlation* (ONO: 291), which is actually a sub-series of the independent series. The task now is to link the space of visual *appearance* to this physical *reality*. Here, of course, Russell invokes the causal theory of perception to correlate the physical objects with the contents of the visual field, thereby producing yet another *series by correlation*. On this, Russell states that:

> . . . according to the general principles which must govern any correspondence of real things with objects of perception, any principle which introduces diversity among objects of perception must introduce a corresponding diversity among real things . . . if there is such a correspondence, it must be supposed that diversity in the effects – i.e., the perceived objects – implies diversity in the causes – i.e., the real objects. Hence if I perceive two objects in the field of vision, we must suppose that at least two real objects are concerned in causing my perception.
>
> (RUP: 179–80).

Further correlation results from the ego changing its position in physical space relative to the group of physical objects. This is based on the principle that:

> . . . it is mainly the relative positions of the object and our body that determine what sensations we shall derive from the object.
>
> (*PP*: 30).

Supplementary to the causal theory of perception is, as we found in the last chapter, Russell's commitment to scientific knowledge in general. Thus, e.g., we know from physics that, considered as physical objects, the thunder and lightning are simultaneous. Our knowledge of the speeds of light and sound will prevent us from assuming that the lightning sense-datum and the later thunder sense-datum have different causes (*PP*: 33).

The result of all this is that the Russell of the Platonist phase holds that a close correspondence between appearance and reality can be achieved. Always, however, with the proviso that:

> . . . although the *relations* of physical objects have all sorts of knowable properties, derived from their correspondence with the relations of sense-data, the physical objects themselves remain unknown in their intrinsic nature, so far at least as can be discovered by means of the senses.
>
> (*PP*: 34).

Nor is there a problem in the fact that ". . . an infinite number of points in the geometrical space will have to correspond to a single simple entity in the perceived space" (RUP: 174). As we have found, from an initial *independent series* of points, a *series by correlation* of physical objects is established. From the second series is established a further *series by correlation*, this latter being a series of sense-data. There is no difficulty in the fact that the first series is compact and continuous, whereas the third one is finite. As Russell showed in his technical work on series, there are contained within compact series (such as govern Newtonian points) sub-series called *progressions*.[24] Progressions "are the very essence of discreteness" (*PoM*: 299), with the natural numbers providing a clear example. The simple finite series that govern sense-data are obviously contained in progressions. So by two steps of correlation we move from a compact series to a finite one.

This, in outline, is how Russell used space and time to set about bridging appearance and reality in the Platonist phase. We might wish that he had said a bit more on some of the detail,[25] but enough has been done, I trust, to vindicate the claim of the last chapter that Russell explicitly set out to answer Kant. I will conclude this section by pointing out that each of the three sorts of "specially important" relations that Russell and Whitehead proclaimed in *Principia Mathematica*[26] figure in this bridging of appearance and reality. The role of relations that generate series is obvious. One-many relations are needed to correlate a progression with a compact series, while one-one relations would be prominent in, e.g., correlating the contents of visual fields resulting from the ego moving through physical space to various locations.

10.3 Space and Time Relations as Exemplars of Relations

In the previous chapter's preview of Russell's various phases and the principles underlying them, two fairly strong claims were made regarding space and time relations. Firstly, space and time relations were claimed to play the central role of being *exemplars* of relations. Hence any change in Russell's theories of space and time would affect his views on relations. Secondly, it was claimed that space and time relations were crucial to Russell's constant preoccupation with bridging appearance and reality. Hence, once again, any change in his theories of space and time would affect his views on relating appearance and reality. So far in this chapter on the Platonist phase we have dealt with the second of these. We turn now to the first claim and consider how distinctive features of space and time relations in the Platonist phase carry over to relations in general. If this claim is substantiated then, obviously, it will add a further dimension to the findings of Part One of this book, viz. that relations play a vital role in much Russellian analysis.

It is no surprise that space and time relations are so central to Russell's philosophy. After all, as we have shown in Chapter Nine, he was reacting to Kant, a classic case of a philosophy being shaped by theories of space and time, whilst, according to Russell, wrongly treating relations as purely subjective. Likewise, Russell's detailed studies of the philosophies of Leibniz (*PL*) and Bergson (POB) provided further examples of philosophical positions being largely shaped by doctrines of space and time. In the case of Leibniz, Russell found that his "... theory of space ... is more or less involved in everything that can be said about his philosophy. ... " (*PL*: 112). (This in no way conflicts with Russell's better known claim that Leibniz's philosophy is "almost entirely derived from his logic" (*PL*: v), in particular from the doctrine that every proposition is of the subject-predicate form. This is so because, as Russell shows, Leibniz's theory of space springs directly from the same doctrine.) The implication of all this, then, is that arriving at sound views on space and time is a prerequisite for developing a new philosophical position in opposition to Kant. From all three philosophers Russell drew the conclusion that space and time *relations* had been poorly understood.

Why is it that spatial and temporal relations are *exemplars* of relations in general for Russell? Not only are they the prime examples of external relations, but they have other features that make them "specially important" amongst relations. Thus, e.g., relational propositions do not in general have a simple correspondence to the facts they represent, since the linguistic symbol for a relation is not itself a relation, i.e. '*A* is to the left of *B*' consists of three terms in a triadic relation (treating 'is-to-the-left-of' as one term), whereas the corresponding fact consists of two terms in a dyadic relation. A "few specially important relations might have been symbolised by relations between words", thus for spatial and temporal relations the proposition and the fact can be made to have a simple correspondence, e.g. '*AB*' can represent the fact that *A* is to the left of *B* (OPWH: 303).

What, then, are the features of spatial and temporal relations that serve as exemplars in the Platonist phase? Consider the basic unanalysable space and time relations of *in* or *at* that we found to hold between particles and points or instants. This notion of an unanalysable dyadic relation between entities becomes the model for all relations in the Platonist phase, with Russell needing compelling reasons before moving to anything more complex. Consider, e.g., the two fundamental cognitive (or epistemic) relations that Russell postulates in the Platonist phase – *acquaintance* and *judgement*.

Acquaintance is "a two-term relation of a subject . . . to a single (simple or complex) object . . . " (NSD: 184). "I am *acquainted* with an object when I have a direct cognitive relation to that object, i.e. when I am directly aware of the object itself" (KAKD: 148).[27] In the Platonist phase, the subject or ego is an irreducible non-material particular located in space and time. The relation of acquaintance between the ego and some other entity is an *external* relation (NSD: 186), and the range of entities that can form the second term of the relation is diverse. Firstly the ego can be acquainted with *sense-data*, which are complex particulars located near the region occupied by the ego, e.g. in the nervous system.[28] In the Platonist phase Russell was coy about whether sense-data are mental or not (NSD: 186),[29] however we will find that by the Empiricist phase he was definite that they are physical objects. Secondly, the ego can be acquainted with *universals*.[30] Besides acquaintance with (say) a particular yellow sense-datum, there is also acquaintance with the universal property *yellow*. Likewise, besides acquaintance with instances of (say) particulars in the before and after relation, there is also acquaintance with the universal relation *before and after*. None of this is incompatible, as far as I can tell, with the asymmetry noted earlier in this chapter between properties and relations: whereas properties are particularised in their instances, relations are not. However, notice that this second case of acquaintance involves direct acquaintance with the world of being. So egos are acquainted with both being and existence.

Why not use egos to relate being and existence, rather than points and instants as described above? In Table 10.1 above are listed three features of points and instants that are *being-like*, thereby making these particular *existents* peculiarly suited to linking the two worlds. Egos, in fact have none of the three features: they are in space and time, they are not changeless (I take it that Russell's Platonism doesn't extend to the ego surviving the death of the material body), and they do not fall under independent series. In addition, egos only have access by acquaintance to the appearance sector of the world of existence, though, of course, they can have knowledge by description of the remainder. So once again the centrality of space and time emerges.

A third kind of acquaintance involves *memory*. In accordance with Pears' dictum that "Russell's theory of mind at this period was certainly designed to achieve maximum exposure to the external world . . . " (1967: 198), the theory of memory in the Platonist phase has the ego directly acquainted with particulars in the past, rather than with *images* of those particulars. In a detailed study of the development of Russell's theories of memory, Pears (1974) dubs the direct acquaintance account "Theory I" and maintains that Russell held it, with some modification from 1912 until 1919, when he replaced it with "Theory II" in which images play an essential role. The changes that "Theory I" underwent from 1912 onwards are what interest us here. As Pears sees it, Russell in the 1912 *Problems of Philosophy* postulates acquaintance with particulars perceived in the remote past. Pears describes this direct acquaintance with past particulars as "paradigmatic remote memory" and likens it to "perception through a time-telescope" (1974: 123). However, by 1915, Pears argues, Russell had changed his mind "about the range of acquaintance with particulars perceived in the past", adopting the less radical view that such acquaintance "does not reach back beyond the limit of the specious present" (1974: 119).[31] Although Urmson (1969) has argued that *Problems of Philosophy* also ought to be seen as restricting acquaintance by memory to the specious present, the texts clearly support the Pears account. What is more, the change pointed out by Pears is only to be expected given the changes in Russell's space and time doctrines in the interim.

As we found earlier in this chapter, in 1912 Russell regarded time as a series of actual instants with particulars having irreducible *at* relations to the instants. Once we allow egos with *at* relations to later instants to have a direct acquaintance relation to particulars *at* earlier instants, it is no surprise that Russell (as Pears puts it) "found no sharp break at the point of transition from immediate to remote memory" (1974: 125). It is precisely because every instant, for Russell in the Platonist phase, is equally a part of the ultimate furniture of the world, that once we posit a basic dyadic relation between egos and particulars at other instants then, metaphysically at least, there is no difference between the recent and the remote. I suggest that this sort of consideration underlies Russell's rather hopeful conclusion in *Problems of Philosophy* that "cases of fallacious memory can probably all be . . . shown to be not cases of memory in the strict sense at all" (*PP*: 117). On the other hand, as we will find in the next chapter, the 1915 restriction of memory acquaintance to the specious present is equally predictable from Russell's theory of time at that period. The key point is that the time series is now constructed from sense-data and their relations, with immediate memory sufficing for the construction. Instants no longer have any metaphysical ultimacy.

A fourth kind of acquaintance in the Platonist phase is acquaintance by *introspection* which covers self-consciousness. While Russell is definite that we are acquainted with such complexes as 'Self-acquainted-with-sense-datum', he is less certain that we are acquainted with the self (*PP*: 49–51).

The influence of Russell's understanding of spatial and temporal relations in the Platonist phase on his understanding of the fundamental irreducible dyadic relation of acquaintance is by now evident. We now consider a similar influence on the other fundamental cognitive relation – *judgment*.

This is not the place to undertake a detailed account of the development of Russell's theories of judgment (or belief).[32] Rather, I want to use this development as an illustration of my claim that Russell took space and time relations to be exemplars of relations in general. What we find is that Russell's early theory of judgment viewed it as an unanalysable dyadic relation between an ego and a proposition, on the model of the unanalysable *at* relation between particles and points and instants. So, in 1904 Russell held that

> . . . there are, apart from and independently of judgment, true and false propositions, and . . . either kind may be assumed, believed or disbelieved. (MTCA: 74).

On this view, judgment, whether correct or incorrect, is a basic dyadic relation between an ego and a proposition. However Russell soon found compelling reasons to discard this simple dyadic model of judgment, largely stemming from his commitment to a correspondence theory of truth.[33]

Hence, Russell was led to formulate his multiple relation theory of judgment. This theory went through several versions, but what is of special interest to us is that all of the versions advocated in the Platonist phase treat the multiple relation complex as having the same logical form as certain spatial or temporal complexes. If judgment or belief is not a dyadic relation, Russell proposed to treat it as a polyadic relation, i.e. as complications of the basic dyadic kind of relation. Take, for instance, Othello's false belief that Desdemona loves Cassio. In the Platonist phase Russell construed this belief as a tetradic relation governing a complex in which Desdemona, the relation *loves* and Cassio are each separately related to Othello.[34] Presumably, a spatio-temporal example having the same logical form might be one that construes triple simultaneity as a tetradic relation governing a complex in which three distinct events are each separately in *at* relations to a given instant. In any case, the key point is that in the Platonist phase Russell definitely saw all relations as mere complications of the fundamental spatio-temporal dyadic relation. This is why Russell was so stunned by Wittgenstein's discovery that ". . . nothing that occurs in space is of the same form as belief" (PLA: 199). But this is a matter that must be taken up in the next chapter.

NOTES

1. *EFG* outlines, e.g., the various contradictions in geometry. See Griffin, 1991: Section 4.8, for discussion.
2. See also PIML: 39.
3. At the start of the Platonist phase, all *existents* had *being* as well. With the developments discussed in Chapter Four, Russell came to talk of the two worlds as *opposites*. The difference appears to be terminological only with the later usage being used in reviews of Meinong to fit Meinong's own usage. In any case, the main points that follow are independent of this trend.
4. For interesting discussion of the limitations of Russell's conception of pure mathematics see Grattan-Guinness, 1984 and 1990.
5. The reader of *Principles of Mathematics* is referred several times to the latter paper. As with *Principles of Mathematics*, PTS is more concerned with the *being* of space and time, but goes further in its discussion of their existence.
6. See also PTS: 276.
7. See also PTS: 276.
8. This is the outcome of PTS, repeated in *Principles of Mathematics* and assumed in *Problems*

of Philosophy.

9. See also PTS: 276.
10. Russell adds that it "seems to be a fact that there are no terms which occupy points but not instants."
11. Later in RUP (p. 180), Russell flirts with the view that mental qualities are in space as well as time. At best he is uncertain on this point.
12. See also *PoM*: 262. Other examples of independent series discussed by Russell in ONO include the colours of the spectrum, the pitches of sounds, magnitudes and wholes and parts. The claim made in the ONO passage is somewhat puzzling. Surely a series S can be ordered without the order being intrinsic. In which case order by correlation to S can be achieved, so long as S is ordered. (Cf. Russell's claim (*IMP*: 30, discussed in Section 7.3) that the integers, like the stars, have no intrinsic order.)
13. This asymmetry has its origins in *Principles of Mathematics*, is clear in *Problems of Philosophy*, Ch. IX, "The World of Universals" and is elaborated in fine detail in the 1912 paper RUP.
14. "Relations, obviously, do not exist anywhere in space. . . . General qualities, such as whiteness, on the contrary, may be said to be in many places at once . . . " (RUP: 168) and " . . . relations do not have instances . . . " (*PoM*: 52). See also MTCA: 48ff. Russell's reasons for holding this view are given in *Principles of Mathematics* and RTC but are not important here. For an account of this see Foster, 1983. Russell dropped this idea of an asymmetry between properties and relations at the end of the Platonist phase. An otherwise interesting article by Winslade (1970) is marred by the assumption that Russell was *always* committed to relations being not particularised. It is no surprise then that Winslade finds Russell's views to be inconsistent.
15. Chapter Seven above.
16. Nicholas Griffin has suggested another possible explanation for Russell claiming that space is "essentially a class of points". It might be that he was thinking of the absolute space of pure mathematics as merely a set of points which could be ordered by a variety of relations. In contrast, the existent physical space would be a set of points on some particular ordering relation.
17. See *PoM*, Ch. L, "The Continuity of Space".
18. *MPD*: 42, spells out this point.
19. See also MTT: 139–40.
20. See, e.g., *PP*: 33.
21. PIML: 34, dating from same year as *Problems of Philosophy*, is explicitly committed to "indivisible" points and instants as existents. Re-reading *Problems of Philosophy* almost fifty years later, Russell himself interpreted it as committed to "points, instants and particles as part of the raw material of the world" (*MPD*: 77).
22. Lowe claims (pp. 3–4) that the projected Vol. IV of *Principia Mathematica*, which was to extend its methods to geometry, was as yet only an idea.
23. See also *OKEW2*: 7–8; SMP: 69; RSDP: 12; and *PoM*, 1937 Introduction to the Second Edition: xi.
24. LOR: 341ff. and *PoM*: 262, 283, 297ff. and 363–4.
25. The works written in 1911 and published in 1912, were, as we have seen, only the start. But then, thanks to Whitehead, Russell moved on to the Empiricist phase.
26. See Chapter Five above.
27. See also *PP*: 46ff.
28. For an influential discussion of the difficulties in Russell's theory of sense-data see Pears, 1967: Ch. III.
29. In *PP*: 137, "mental facts" are evidently a separate category from "facts concerning sense-data" though both are "private".
30. See, e.g., KAKD: 150–1 and *PP*: 52.
31. Main texts are SAI and OET.
32. A good account is provided by Griffin (1985).

33. For details see Griffin (1985).
34. Griffin (1985) outlines 1910 and 1912 versions that involve this scheme. Their differences are not important to the matter at hand. Griffin also mentions a 1906 version that *doesn't* involve this scheme. However Russell only put this version forward as a possible theory – he never adopted it.

CHAPTER 11

Space and Time in the Empiricist Phase (1914–1918)

In many ways Russell's theories of space and time in the Empiricist phase are much simpler than their counterparts in the Platonist phase. The being/existence distinction is scrapped, and, though some of its features survive in the doctrine that existents are of many different *types*, this does not significantly affect the project of bridging appearance and reality. Hence some of the complications of the Platonist phase are avoided. For our present concerns, the main point about the Empiricist phase is that it has space and time in the world of reality (physics) as logically dependent on their counterparts in the world of appearance, thereby reversing the logical order of the Platonist phase. This entails that sense-data, sensibilia, and their relations will be the basic entities reached by analysis, rather than irreducible *at* or *in* relations to absolute points and instants.

11.1 Bridging Physics and Sense-data

As discussion in previous chapters has shown, it was Whitehead's idea that points and instants could be logically constructed from sense-data that initiated the Empiricist phase for Russell.[1] As we saw, Russell "was delighted with this fresh application of the methods of mathematical logic" since it "seemed to suggest that all the smoothness of the concepts used in theoretical physics could be attributed to the ingenuity of mathematicians rather than to the nature of the world" (*MPD*: 77). Note, however, that Russell is *not* here abandoning his acceptance of scientific knowledge. Instead, in undertaking "to build a bridge from physics to sense-data" (*OKEW2*: 109), he is attempting an *analysis* of basic scientific entities rather than accepting them at face value, as had been the tendency in the Platonist phase. So, rather than treat sense-data in such a way that there is merely, no clash with the claims of science, interpreted realistically, the aim is to try to *interpret* science in terms of sense-data thereby anticipating its next stage of development. Such an aim means that from the Empiricist phase onwards we often find Russell expressing sentiments like the following:

> . . . you may say that the whole of our problem belongs rather to science

than to philosophy. I think perhaps that is true, but I believe the only difference between science and philosophy is, that science is what you more or less know and philosophy is what you do not know. Philosophy is that part of science which at present people choose to have opinions about, but which they have no knowledge about. Therefore every advance in knowledge robs philosophy of some problems which formerly it had, and if there is any truth, if there is any value in the kind of procedure of mathematical logic, it will follow that a number of problems which had belonged to philosophy will have ceased to belong to philosophy and will belong to science.

(PLA: 243).

Russell's trust in science is graphically illustrated by one crucial respect in which there is *no change* between the Platonist and Empiricist phases. In both cases Russell is convinced that the world is *Newtonian*. The change is simply that points, instants and particles are no longer to be taken as irreducible entities – instead they will be logical constructions of sense-data and sensibilia. This means that physical space and time are now *relative* rather than absolute (as in the Platonist phase). As already suggested in Table 9.1 above, in the Empiricist phase, bridging appearance and reality will involve correlating a perceptual space that is *relative* and *subjective* with a physical space that is *relative* and *objective*. How this is done will be discussed shortly.

Einsteinian relativity was, of course, becoming well known during Russell's Empiricist phase and one might conjecture as to why he did not realise its full implications for his philosophy earlier than 1919. Certainly in the academic year starting in mid-1913 Russell lectured on the theory of relativity (Wiener, 1964: 191), yet in his major book in the following year he wrote as follows:

There is supposed to be one all-embracing space in which the motion takes place, and until lately we might have assumed one all-embracing time also. But the principle of relativity has given prominence to the conception of 'local time', and has somewhat diminished men's confidence in the one even-flowing stream of time. Without dogmatising as to the ultimate outcome of the principle of relativity, however, we may safely say, I think, that it does not destroy the possibility of correlating different local times, and *does not therefore have such far-reaching philosophical consequences as is sometimes supposed*. In fact, in spite of difficulties as to measurement, the one all-embracing time still, I think, underlies all that physics has to say about motion. We thus have still in physics, as we had in Newton's time, a set of indestructible entities which may be called particles, moving relatively to each other in a single space and a single time.

(*OKEW1* (i.e. 1914 Edition): 103–4, my italics).

Not surprisingly, Russell removed this and similar passages from his 1926 and 1929 revisions of *Our Knowledge of the External World*. If Wiener is correct, it seems that the Russell of 1913 welcomed Einsteinian relativity for the support it offered to the "new emphasis on the observer" (Wiener, 1964: 191), a characteristic feature of his Empiricist phase, without being aware that it would undermine his whole enterprise. Clearly, then, Russell had some years of familiarity with relativity theory before the decisive events of 1919 that opened his eyes to its full significance (*Auto.*: 328).[2]

So, space and time in both the Platonist and Empiricist phases are assumed to be strictly Newtonian, with the difference between the two cases being a reversal of the order of logical dependence. In the Platonist phase, the *basic* Newtonian spatial and temporal relations, as shown in the previous chapter, held between the changeless points and instants that constituted absolute space and time. The simpler spatial and temporal relations of the world of appearance were found to derive from the more complex basic relations by two steps of correlation of series. In the Empiricist phase, physical space and time are relative because the basic relations now hold between sense-data, with the Newtonian space and time of physics being established by correlation and construction.

Starting, then, from appearance, how does Russell make the bridge to the world of physics? As before, he notes that the "different senses have different spaces. The space of sight is quite different from the space of touch: it is only by experience in infancy that we learn to correlate them" (*OKEW1*: 113).[3] Also, as before, Russell's discussion concentrates on *visual* space. Crucial to the bridging of the two worlds is the famous theory "that space has six dimensions and not only three" (*MPD*: 79). Russell's first step is to postulate an infinity of *perspectives*, of which actual visual spaces are a subset:

> Let us imagine that each mind looks out upon the world, as in Leibniz's monadology, from a point of view peculiar to itself; and for the sake of simplicity let us confine ourselves to the sense of sight, ignoring minds which are devoid of this sense. Each mind sees at each moment an immensely complex three-dimensional world; but there is absolutely nothing which is seen by two minds simultaneously. When we say that two people see the same thing, we always find that, owing to difference of point of view, there are differences, however slight, between their immediate sensible objects . . . The three dimensional world seen by one mind therefore contains no place in common with that seen by another, for places can only be constituted by the things in or around them. Hence we may suppose, in spite of the differences between the different worlds, that each exists entire exactly as it is perceived, and might be exactly as it is even if it were not perceived. We may further suppose that there are an infinite number of such worlds which are in fact unperceived. If two men are sitting in a room, two somewhat similar worlds are perceived by them; if a third man enters and sits between them, a third world, intermediate between the two previous worlds, begins to be perceived. It is true that we cannot reasonably suppose just this world to have existed before, because it is conditioned by the sense-organs, nerves, and brain of the newly arrived man; but we can reasonably suppose that *some* aspect of the universe existed from that point of view, though no one was perceiving it. The system consisting of all views of the universe perceived and unperceived, I shall call the system of 'perspectives'; I shall confine the expression 'private worlds' to such views of the

> universe as are actually perceived. Thus a 'private world' is a perceived 'perspective'; but there may be any number of unperceived perspectives. (*OKEW1*: 87–8).[4]

So far, then, we have a plurality of three-dimensional spatial worlds some of which are actually perceived. The next step is to establish that there is a *further space* consisting of relations between the perspectives:

> . . . it is possible, sometimes, to establish a correlation by similarity between a great many of the things of one perspective, and a great many of the things of another. In case the similarity is very great, we say the points of view of the two perspectives are near together in space; but *this space in which they are near together is totally different from the spaces in the two perspectives.* It is a *relation between the perspectives*, and is not in either of them; no one can perceive it, and *if it is to be known it can be only by inference*. Between two perceived perspectives which are similar, we can imagine a whole series of other perspectives, some at least unperceived, and such that between any two, however similar, there are others still more similar. In this way the space which consists of relations between perspectives can be rendered continuous, and (if we choose) three-dimensional. (*OKEW1*: 88–9, my italics).

This gives us a plurality of three-dimensional perspectives which are themselves elements of a single *inferred* three-dimensional space. This then leaves us with the final step which is:

> . . . to explain how the private space of a single perspective is correlated with part of the one all-embracing perspective space. Perspective space is the system of 'points of view' of private spaces (perspectives), or, since 'points of view' have not been defined, we may say it is the system of the private spaces themselves. These private spaces will each count as one point, or at any rate as one element, in perspective space. They are ordered by means of their similarities.
>
> *OKEW1*: 89–90).

Russell goes on to sketch how the ordering by similarities would work. The essential idea is that perspectives are arranged in a three-dimensional series on the basis of features of their contents. So, e.g., take a large number of similar perspectives and concentrate on one of their aspects such as "the appearance of a circular disc, such as would be called a penny" (*OKEW1*: 90). A tentative three-dimensional series can be formed as follows. Those perspectives in which the penny appearance is circular will be in a straight line in the order of the sizes of the circular aspects. On either side of this will be perspectives in which the penny appearance is slightly elliptical, once again ordered according to relative sizes. Features of the perspectives other than penny appearances can be used both to render the ordering into a three-dimensional series more definite and to extend the range of the series into perspectives that lack penny appearances. So we end up with a definite three-dimensional series of perspectives into which others would be fitted.

In Chapter Seven criticisms by A.J. Ayer and R.M. Sainsbury of this Russellian procedure were considered and rejected. The basis of the rejection was that Ayer and Sainsbury both fail to realise that what we have here are *series* of perspectives, so that the *relations* between the perspectives are just as important as the individual perspectives. Their arguments fail because they are directed at particulars only. Hence my conclusion is that Russell's idea here is not as implausible as some of his critics have maintained.

Rather than pursue this further, I will draw attention to the way this procedure transfers the characteristics of the space of appearance to the space of reality, i.e. "the one all-embracing perspective space". Basic spatial relations are given in the world of appearance, i.e. within individual perspectives. Examples would be relations between the penny appearance and (say) the table on which it appears to rest. These perceived relations between the contents of individual three-dimensional perspectives are the same *sorts* of spatial relations that are *inferred* to hold between perspectives. This is so because the *correlation* of perspectives centres on basic Euclidean geometrical assumptions of distance-size relationships, angle-point of view relationships, etc. The correlated three-dimensional perspectives are then taken to be points in the inferred three-dimensional perspective space, hence Russell's claim that space has six dimensions rather than three. Points can be logically constructed from the sense-data and their relations within a given perspective along the lines sketched by Russell (*OKEW1*: 114–5).[5] This lends plausibility to the idea of treating a perspective as a point and enables the inferred three-dimensional perspective space to be aligned with the Newtonian space of physics. Note, however that the former is an *inferred* space consisting of relations between perspectives (perceived and unperceived, i.e. inferred), whereas the latter is a *constructed* space, since its points are logical constructions.

Presumably the same kinds of considerations apply to other spaces such as the space of touch. I take it that there is one all-embracing touch space parallel to its perspective counterpart. The individual ego learns to correlate the space of sight with the space of touch, but the "one space into which both kinds of sensations fit is an intellectual construction, not a datum" (*OKEW1*: 113). This space, in which various sorts of sensation fit, appears *not* to be equivalent to the *inferred* three-dimensional perspective space, though Russell is not as explicit on this as one would like.

The main outcome, then, of Russell's linking of perceived space and physical space is that *both have the same sorts of spatial relations*. It is largely because the spatial relations are of the same sort that Russell is able to retain the Platonist phase conclusion that appearance and reality "more or less correspond" (though, of course, *reality* is quite different in the two cases). As we will find in the next chapter, Russell was forced to abandon this assumption in the Modified Empiricist phase.

As noted above, the Empiricist phase is also marked by Russell's commitment to one all-embracing Newtonian time. So once again the task is to show

how this can be constructed from time in the world of appearance. According to Russell, the

> ... question of time, so long as we confine ourselves to *one private world*, is rather less complicated than that of space, and we can see pretty clearly how it might be dealt with by such methods as we have been considering.
> (*OKEW1*: 116, my italics).

Russell goes on to argue that our

> ... experience provides us with two time-relations among events: they may be simultaneous, or one may be earlier and the other later. These two are both part of the crude data; it is not the case that only the events are given, and their time-order is added by our subjective activity. The time-order, within certain limits, is as much given as the events.
> (*OKEW1*: 116).[6]

Thus, these two basic time relations are given in experience, just as we found the basic spatial relations were experiential. Russell proceeds on this basis to show how instants can be constructed with the requisite serial properties for physical time. However, these results are all derived from one private world. The

> ... correlation of the times of *different private worlds* so as to produce the one all-embracing time of physics is a more difficult matter.
> (*OKEW1*: 121, my italics).

However, claims Russell, the difficulties are "technical" rather than philosophical:

> What we call one sound will be heard sooner by people near the source of the sound than by people further from it, and the same applies, though in a less degree, to light. Thus two correlated appearances in different worlds are not necessarily to be regarded as occurring at the same date in physical time ...
> (*OKEW1*: 122).

So the "complicated technical problem" will be to correlate different private times so that "the simplest possible statement of the laws of physics" results. In principle, at least, the philosophical problem of bridging temporal appearance and reality is solved.

11.2 The Causal Theory of Perception

In Chapter Nine it was argued that all of Russell's post-idealist philosophising is an attempt to answer Kant. General principles underlying this attempt included:

(i) a commitment to the general findings of science;

(ii) a conviction that private space and time can be closely related to their objective counterparts; and

(iii) a conviction that the nature of appearance enables us to infer something of the nature of its causal antecedents, particularly their relational characteristics.

So far in this chapter on the Empiricist phase, we have dealt with (i) and (ii). Let us now turn to (iii).

The physical object that had a causal relationship to sense-data in the Platonist phase now becomes a logical construction out of sense-data.

> We can now define the momentary common-sense 'thing', as opposed to its momentary appearances. By the similarity of neighbouring perspectives, many objects in the one can be correlated with objects in the other, namely, with the similar objects. Given an object in one perspective, form the system of all the objects correlated with it in all the perspectives; that system may be identified with the momentary common-sense 'thing'. Thus an aspect of a 'thing' is a member of the system of aspects which *is* the 'thing' at that moment.
>
> (*OKEW1*: 89).

Once the technical problems of correlating different private times are overcome, we can construct the thing-over-time. However at any moment it is a peculiar combination of subjectivity and objectivity:

> All the aspects of a thing are real, whereas the thing is a mere logical construction. It has, however, the merit of being neutral as between different points of view, and of being visible to more than one person, in the only sense in which it can ever be visible, namely, in the sense that each sees one of its aspects.
>
> (*OKEW1*: 89).

This combination of subjectivity and objectivity is used to link psychology and physics and, more importantly for us, underlies Russell's *causal theory* of perception in the Empiricist phase. The crucial idea is that

> . . . *two* places in perspective space are associated with every aspect of a thing: namely, the place where the thing is, and the place which is the perspective of which the aspect in question forms part. Every aspect of a thing is a member of two different classes of aspects, namely: (1) the various aspects of the thing, of which at most one appears in any given perspective; (2) the perspective of which the given aspect is a member, i.e. that in which the thing has the given aspect. The physicist naturally classifies aspects in the first way, the psychologist in the second. The two places associated with a single aspect correspond to the two ways of classifying it. We may distinguish the two places as that *at* which, and that *from* which, the aspect appears. The 'place at which' is the place of the thing to which the aspect belongs; the 'place from which' is the place of the perspective to which the

aspect belongs.

(*OKEW1*: 92).

The two places in the inferred perspective space that are associated with each aspect of a thing are the heart of the causal theory of perception. Since, e.g., Ayer (1972: Ch. 3)[7] has promoted the idea that Russell in his 1914 works abandoned the causal theory of perception in favour of a version of phenomenalism, it is vital that we be clear on how the two places doctrine supports the causal theory.

A causal theory of perception is, in essence, any theory which places sense-data (or their equivalents) at the end of causal chains that extend back to physical objects or events. It is non-controversial that Russell held such a theory in the Platonist and Modified Empiricist phases. If we can establish that the 1914 theory of perception is a causal one we will have shown that the same is true of the Empiricist phase.[8] As Salmon (1974)[9] has pointed out, the difficulty of understanding how a sense-datum can be caused by a physical object, whilst itself being one of the sense-data out of which the object is constructed, is the basis of the view that Russell abandoned the causal theory. This is the difficulty that we must now resolve.

I have referred to the peculiar combination of subjectivity and objectivity in Russell's construction of things. It is vital to be clear about just what is objective. *Both* of the different places in the inferred perspective space are objective, i.e. they are different places in an actual space associated with a sensibile whether or not it becomes a sense-datum. (Of course, when a sensibile becomes a sense-datum there is a *third* place associated with it, a *subjective* place in that it is located in a private visual space.) In saying that two objectively different places in perspective space are associated with every sensibile, it is *not* being claimed that the sensibile is *located* in two different places. The sensibile is located only at the perspective of which it is a member, this perspective being a point in perspective space. This means that the sensibilia or aspects out of which the thing is logically constructed are all at different locations in perspective space. However, as we found above, perspectives can be arranged in a three dimensional series on the basis of features of their contents. We can, according to Russell, use this to define *the place in perspective space where the thing is located*. It works as follows:

> . . . let us again consider the penny which appears in many perspectives. We formed a straight line of perspectives in which the penny looked circular, and we agreed that those in which it looked larger were to be considered as nearer to the penny. We can form another straight line of perspectives in which the penny is seen end-on and looks like a straight line of a certain thickness. These two lines will meet in a certain place in perspective space, i.e. in a certain perspective, which may be defined as 'the place (in perspective space) where the penny is'.
>
> (*OKEW1*: 90–1).

Russell admits that this simple procedure needs further extension, e.g. we can only get so close to the penny before it ceases to present any appearance. He briefly indicates how the method could be extended and insists that

> . . . it is plain that such niceties cannot affect the principle, and can only introduce complications in its application.
>
> (*OKEW1*: 91).

If we accept this procedure, then we have the second objective place in perspective space that is associated with each sensibile. The theory of perception in Russell's 1914 works is a *causal* theory because the features of the aspect that is part of any perspective depend on the distance between the perspective and the place where the 'thing' is, as well as on the other contents of the local perspective space. Russell puts it as follows:

> Let us now endeavour to state the fact that the aspect which a thing presents at a given place is affected by the intervening medium. The aspects of a thing in different perspectives are to be conceived as spreading outwards from the place where the thing is, and undergoing various changes as they get further away from this place. The laws according to which they change cannot be stated if we only take account of the aspects that are near the thing, but require that we should also take account of the things that are at the places from which these aspects appear. This empirical fact can, therefore, be interpreted in terms of our construction.
>
> (*OKEW1*: 92–3).

So, even though the *thing*, being a logical construction out of widely located sensibilia and sense-data, is not an actual individual, the features of the three-dimensional series of sense-data are such that causal chains appear to emanate from a definite place in perspective space. This place is defined as being the place where the thing is, and, hence, is associated with every sense-datum and sensibile that is part of the thing.

Hence the 1914 theory of perception, though different from its predecessor, is still a *causal* theory in the sense defined several pages back. Interestingly the central doctrines that Russell derived from empiricism and rationalism (outlined in Chapter Nine) survive unscathed in the new causal theory of perception. The empiricist doctrine that all qualities are secondary clearly applies, since all sense-data are private to a single percipient. Because Russell in the Empiricist phase treats sense-data as non-persistent, impermanent and transient,[10] it follows that even if I occupy the perspective recently vacated by another ego, its contents will have changed to some extent. In place of the unknowable qualities of the causes of sense-data in the Platonist phase, we have the unsensed sensibilia in the Empiricist phase. Likewise the relational characteristics of perspectives are now correlated with those of other perspectives as their causal antecedents, rather than with physical objects. The doctrine derived from rationalism, that distinguishes private space and time from their objective counterparts, has been central to most of this chapter so far. Once again I trust that the claim that Russell was attempting to answer

Kant is supported by the preceding account of the Empiricist phase.

Salmon (1974) has argued that if Russell's 1914 theory of perception is a causal one, then he was not a phenomenalist. However, given the unusual character of the 1914 theory, I am unsure whether, although a causal theory, it might not be compatible with *some* versions of phenomenalism.[11] However, more important than philosophical labels is clarity about just what Russell's position was. In Chapter Seven it was argued that whether or not Russell's theory was phenomenalist, it was distinguished by the *relations* between sense-data being as ontologically significant as the *sense-data* themselves. A consequence was that some major arguments, commonly directed at certain well-known versions of phenomenalism, failed against Russell's position.

Because I take it that the ontological import of relations for Russell is one of the main theses of this book, as well as one of the most misunderstood aspects of his work, I will pause to suggest that this chapter has provided further evidence for this claim. It has emerged from the above that Russell sharply distinguishes between the *inferred* perspective space and the *constructed* space of Newtonian physics. What I want to argue is that unless one treats relations as ontologically real (as Russell does), then it is hard to maintain the distinction between the two spaces. Indeed, without relations treated in some sense realistically, the ontology of the two spaces is the same, i.e. sense-data and sensibilia. In that case it becomes irresistible to follow Russell's 1914 "supreme maxim in scientific philosophising. . . . *Wherever possible, logical constructions are to be substituted for inferred entities*", thereby elevating the constructed Newtonian space to the centre of philosophical interest. However Russell does not do this. It is absolutely clear that he accords philosophical priority to the one all-embracing perspective space:

> . . . this space . . . is totally different from the spaces inside the . . . perspectives. It is a relation between the perspectives . . . no one can perceive it, and if it is to be known it can be only by inference.
>
> (*OKEW1*: 88).

Russell goes on to describe how the "space which consists of relations between the perspectives can be rendered continuous, and . . . three-dimensional". Being constituted of relations between complex sensibilia and sense-data, this space is *relative* rather than absolute. However it is also an *objective* space:

> . . . there is only one space in which the perspectives themselves are the elements. There are as many private spaces as there are perspectives; there are therefore at least as many as there are percipients, and there may be any number of others which have a merely material existence and are not seen by anyone. But there is only one perspective-space, whose elements are single perspectives, each with its own private space.
>
> (*OKEW1*: 89).

Of course, it is this feature of perspective space that underlies the notion that *two* objective places are associated with each aspect of a thing.

We can describe the objectivity of perspective space another way. For Russell in the Empiricist phase, perspectives are part of the ultimate furniture of the world. Hence the serial relations that perspectives instantiate, though not experienced, are likewise real. So perspective space, though inferred, is real. When we recall that Russell was "delighted" with Whitehead's method of logical constructions because it suggested that "all the smoothness of the concepts used in theoretical physics could be attributed to the ingenuity of mathematicians rather than to the nature of the world" (*MPD*: 77), it is clear that the constructed Newtonian space comes into this category, while perspective space does not because it belongs to "the nature of the world".

As you would expect, Russell's Empiricist phase ontology is more economical than its Platonist counterpart. Reality comprises *particulars* (sensibilia and minds) and *universals* (qualities and relations of various orders). All of these "have in their various ways some kind of reality that does not belong to anything else" (PLA: 234).[12] Appearance, comprising sense-data and universals, is a more representative sub-class of reality than was the case in the Platonist phase. Once again egos are not directly experienced, so that by the end of the Empiricist phase Russell was increasingly attracted by neutral monism, but was yet to be finally converted.[13] The Platonist phase doubts about whether sense-data are physical or mental are decisively resolved in favour of their physical status (RSDP: 8–9). The independent series, which for Russell in the Platonist phase enjoyed "a distinguished place among entities" in the world being, are abolished, along with the world of being. Now there are simply many various relations amongst terms which give rise to various different orderings (*IMP*: 29–30). The three-dimensional serial ordering of perspectives would seem to be the closest we come to an independent series of entities in the Empiricist phase. Likewise the irreducible *at* or *in* relations of particles to points and instants are no more. As we will find shortly, *acquaintance* becomes the prototype of the basic dyadic relation. Once again the three sorts of "specially important" relations, that Russell and Whitehead stressed in *Principia Mathematica*, figure prominently in the bridging of appearance and reality. One-one relations between the contents of two perspectives are used, e.g., to establish that they are close together in perspective space. Perspectives themselves fall under relations that generate series, while one-many relations govern, e.g., things and their aspects.

11.3 Space and Time Relations as Exemplars of Relations

It remains in this chapter to investigate the extent to which space and time relations in the Empiricist phase continue to serve as exemplars of relations in general. In Chapter Ten we found a number of reasons why Russell was tempted to see them this way, and found that this did indeed carry through to his understanding of some typical non-spatio-temporal relations. We also found that Russell in the Platonist phase saw science as postulating a set of indestructible particles moving relative to one another throughout the indestructible points and instants of a single space and a single time. This led, we found, to the dyadic *at* or *in* relations of particles to points and instants being viewed as the *paradigm* relations between existents. Likewise in the world of being, the serial relations between the points (or instants) were paradigms of independent series. In the Empiricist phase, the once indestructible particles, points and instants have become logical constructions. Clearly, they can no longer serve as the relata of basic dyadic relations, nor as the terms of the most fundamental serial relations. Nevertheless, as was shown at the start of this chapter, Russell in the Empiricist phase remained committed to the Newtonian picture of the world, i.e. persistent entities moving in an all-embracing space and time. Likewise, Russell's general philosophical commitment to pluralism requires him to posit a multitude of particulars related to one another by external relations. So it is no surprise that once again a basic dyadic relation becomes central to philosophical analysis. In this case it is the relation of *acquaintance.*

The reasons for this are fairly clear. As was shown in Section 4.4, the premisses reached by logical analysis for Russell's Empiricist phase always included premisses about inferred entities, viz. other minds and sensibilia. This meant that primitive epistemological propositions were, at best, only part of the premisses reached by logical analysis. Nevertheless, the characteristic feature of the Empiricist phase was the principle that where possible logical constructions were to be substituted for inferred entities. For Russell a logical construction is more secure epistemically than an inferred entity because all of the elements of the construction are objects of acquaintance. This means that the relation of acquaintance, which had already figured in Russell's work in the Platonist phase, assumes a more central role in the Empiricist phase. In particular, Russellian analysis in the Empiricist phase will proliferate logical constructions at the expense of inferred entities. Hence, the relation of acquaintance determines what counts as an element suitable for inclusion in a logical construction, thereby underpinning analysis in this phase.

So the basic dyadic relation of the Empiricist phase is *acquaintance*:

> . . . acquaintance is a dual relation between a subject and an object which need not have any community of nature. The subject is 'mental', the object is not known to be mental except in introspection. The object may be in the present, in the past, or not in time at all; it may be a sensible particular, or a universal, or an abstract logical fact. All cognitive relations – attention, sensation, memory, imagination, believing, disbelieving, etc. – presuppose acquaintance.
>
> (ONA: 127).

Russell's continuing adherence to dualism is evident.[14] Also noteworthy is the fact that there are *three* different types of acquaintance corresponding to the three different types of objects with which subjects have acquaintance.[15]

At first sight the basic relation of *acquaintance* might appear to have little in common with the dyadic *at* or *in* relation of the Platonist phase. However, a closer examination will reveal quite a few similarities. As we have seen, acquaintance is a relation between a relatively perduring particular (a subject or ego) and an object whose nature varies. Consider first the subject or ego. At the start of the Modified Empiricist phase, when Russell had abandoned dualism, thereby dispensing with the subject or ego, he commented as follows:

> The subject . . . appears to be a logical fiction, like mathematical points and instants. It is introduced, not because observation reveals it, but because it is linguistically convenient and apparently demanded by grammar. Nominal entities of this sort may or may not exist, but there is no good ground for assuming that they do. The functions that they appear to perform can always be performed by classes or series or other logical constructions, consisting of less dubious entities. If we are to avoid a perfectly gratuitous assumption, we must dispense with the subject as one of the actual ingredients of the world.
>
> (*AMi*: 141–2).[16]

If the subject or ego is really in the same category as points and instants, why didn't Russell dispense with it in the Empiricist phase? The answer, I suggest, stems from acquaintance being the best candidate for a basic relation on which to build a version of empiricism of the kind favoured by Russell. It seems that Russell thought that he needed a basic relation between the ego and sense-data in order to establish the relations between perspectives, and, hence, to correlate appearance and reality. The ego as a relatively perduring particular successively has relations of acquaintance to various perspectives whose sense-data are, of course, transient. The perspectives are ordered in the manner described earlier in this chapter, thereby enabling perspective space to be inferred. Russell later asserted that the dualism of mind and matter depends "upon the belief that the space of physics can be identified with the space of sensible experience" (*OP*: 192). This claim is consonant with his own philosophical development in which he abandoned dualism for neutral monism only when he came to realise that the inferred perspective space has no close connection with the space-time of Einsteinian physics. As we will find in the next chapter, the Modified Empiricist phase is marked by the view that, at best, private space and time have only a partial and indirect correlation with any ultimate physical reality. All of this provides an answer to the puzzle of why Russell, though attracted to neutral monism virtually throughout the Empiricist phase, only adopted it in 1919 when he realised the full significance of Einstein's work for his own philosophy. Once again we have an illustration of the thesis that space and time theories shape Russell's overall philosophical views.

Having dealt with the ego, what about the other term in the relation of acquaintance, i.e. the object with which the ego is acquainted? As already noted, the object is of three different types: sensible particulars, universals and abstract logical facts. The case of the ego acquainted with sensible particulars is the one that takes on features that most closely resemble the dyadic *at* or *in* relations of the Platonist phase. In *perception* the ego is presently acquainted with a momentary three-dimensional perspective and (as discussion earlier in this chapter showed), Russell held each such three-dimensional perspective to be a point (or minimal volume) in perspective space. As we found, perspective space is an inferred space generated by the *serial relations* between perspectives, where each three-dimensional perspective is a point or minimal volume in a three-dimensional perspective space. The ordering of perspectives is possible because, in effect, they are points or minimal volumes with properties, i.e. the properties of each perspective bear serial relations to the properties of neighbouring perspectives both across space and time. This means that in perception the ego acquainted with a three-dimensional perspective in *private space*, actually has, from the viewpoint of *perspective space*, an *at* relation to a *point* (or minimal volume). Thus both the dyadic *at* relation and the serial relations that were fundamental to space and time in the Platonist phase, are retained in the Empiricist phase. Acquaintance, the fundamental relation of the Empiricist phase, is *space-like* in the way that Russell of the Platonist phase thought of all relations being space-like.

Memory is, of course, the case where the ego is acquainted with a *past* sensible particular. As was noted towards the end of the previous chapter, up to 1919 Russell viewed memory as the ego having direct acquaintance with particulars in the past. However Russell found that in logically constructing the Newtonian time series he only required immediate memory, i.e. acquaintance with "an object which has recently been a sense-datum, but is now felt as past, though still given in acquaintance" (OET: 225).[17] Accordingly Russell no longer needed *remote* memory as a case of direct acquaintance and dropped it. So acquaintance, the Empiricist phase's fundamental relation, is *time-like* as well as *space-like*, showing yet again the space and time influence on relations in general.

Let us now turn to universals, the second type of object of acquaintance. In the Platonist phase we found that universals, specifically relations, were crucial to connecting the worlds of being and existence, as well as, of course, the worlds of appearance and reality. This led to Russell postulating an asymmetry between relations and properties in which only the latter are particularised. With the being/existence distinction discarded in the Empiricist phase, a different view of universals is obviously required. This Russell certainly provides, though one would like a few more of the details to be spelt out. Perhaps Russell didn't have a chance to write more on universals before the relatively brief Empiricist phase came to its sudden end for the reasons detailed in the next chapter. In any case it is clear that Russell holds that there are both properties and relations and we are acquainted with them.[18] Russell's rejection of any asymmetry between properties and relations *qua* universals is evident from his classifying properties as "monadic relations" (PLA: 177). Then there is the point that universals comprise different logical types (PLA: 234).[19] In Chapter Eight, I discussed the difficulties this poses for *talk* about universals. For Russell in the Empiricist phase, to say that universals exist is simply to say that certain propositional functions are "capable of truth" (PLA: 225); (an example would be 'x is a colour between blue and yellow').

It remains to consider abstract logical facts as the third kind of object with which egos are acquainted. This sort of acquaintance only makes its appearance in Russell's 1913 *Theory of Knowledge* manuscript and the 1914 papers that he published from it. It is a peculiar sort of acquaintance because, according to Russell,

> . . . logical objects cannot be regarded as 'entities' . . . therefore, what we shall call 'acquaintance' with them cannot really be a dual relation. The difficulties which result are very formidable. . . .
>
> (*Papers* 7: 97).

Some idea of the difficulties can be gauged from the contrast between this last quotation and the following one from the opening paragraph of the *same* work (*Theory of Knowledge*):

> It will be maintained that acquaintance is a dual relation between a subject and an object . . . The subject is 'mental', the object . . . may be a sensible particular, or a universal, or an abstract logical fact.[20]

This is not the occasion to sort out the intricacies of this theory.[21] Rather, the *reasons* why Russell introduced acquaintance with logical facts are of interest. As was detailed in the previous chapter, Russell in the Platonist phase attempted to explain judgment as a space-like relation. Initially judgment was viewed as a basic dyadic relation, then later as a multiple relation. In late 1912 and early 1913, Wittgenstein directed a number of criticisms at Russell's multiple relation theory of judgment. Part of Russell's effort to overcome the objections was to introduce the logical form as a term in the judgment complex. This led to difficulties about whether or not logical forms are entities (as the above contradictory quotations from *Theory of Knowledge* illustrate). As it turned out, Wittgenstein's sustained criticism was fatal to Russell's multiple relation theory in all its versions, so that Russell never completed *Theory of Knowledge*. This was the end of Russell's attempts to account for judgment as a space-like relation. Hence Russell's 1918 acknowledgement of Wittgenstein's discovery:

> . . . nothing that occurs in space is of the same form as belief. I have got on here to a new sort of thing, a new beast for our zoo, not another member of our former species but a new species. The discovery of this fact is due to Mr Wittgenstein.
>
> (PLA: 199).

So we find that, in the Empiricist phase, the notion of space and time relations as exemplars of relations in general starts to break down. Our account of the three types of acquaintance relation has shown that only the one involved with particulars is space-like and time-like. This is a trend which we will see gathering even more momentum in the Modified Empiricist phase to which we now turn.

NOTES

1. The logical construction from sense-data of physical objects, like tables, may have been Russell's idea. See Lowe, 1974: 9.
2. For discussion of the effects of the theory of relativity on Russell see Wood, 1957: 132–3. Presumably the gaps in Russell's knowledge of relativity in 1913 included Minkowski's contribution, which was first presented in September 1908. Another possibility is that Russell knew of the Minkowski contribution in 1913, but wasn't convinced of its value until the empirical confirmation of the general theory of relativity in 1919.
3. See also RSDP: 13.
4. Essentially the same account is given in RSDP and UCM.
5. This section is revised in *OKEW2* to incorporate improved methods of construction published by Whitehead in 1919 and 1920.
6. A more detailed treatment is given in OET.
7. Sainsbury (1979: 201) appears to agree with Ayer by denying that Russell has a causal theory between 1914 and the mid-1920's.
8. PLA (224ff. and Lecture VIII) show that in 1918 Russell still adhered to the 1914 doctrines on perception. Though some commentators distinguish Russell's 1914 "phenomenalism" from the 1918 "logical atomism", it does not affect my Empiricist phase since the space and time theories stay the same.
9. Salmon's discussion of the 1914 theory is very helpful and her conclusion that it is a *causal* theory appears to be convincing.
10. See, e.g., *OKEW1*: 104 and RSDP: 8–9.
11. Certainly, if Russell was a phenomenalist in this period, his form of phenomenalism was an unusual one, since it is absolutely clear that the sense-data into which he analysed physical objects were physical and not mental.
12. *Logical forms* are another possible candidate for inclusion amongst entities. However they are only considered for this category in the 1913 *Theory of Knowledge* (*Papers* 7) which Russell abandoned, so I omit them here. For discussion see Griffin, 1980 (especially Section 9). See also Griffin (1985).
13. See the discussion of this in Chapter Four above.
14. See the discussion of the Empiricist phase dualism in Chapter Four above.
15. For discussion of this see Griffin, 1980 and 1985.
16. See also OPWH: 294.
17. The construction of the time series is sketched in OET.
18. See, e.g., *Papers* 7: 95–6. (This work comes at the borderline between the Platonist and Empiricist phases.)
19. See also LA: 169
20. This, is, of course, part of an earlier quotation in this chapter from ONA. ONA was part of the early *Papers* 7 material that Russell himself published in 1914.
21. Once again, the reader is referred to Griffin, 1980 and 1985, for detailed critical discussion.

CHAPTER 12

Space and Time in the Modified Empiricist Phase (1919 Onwards)

We found in the previous chapter that while the shift from Platonism to Empiricism was in many respects a drastic one, in the vital business of bridging appearance and reality Russell retained the same fundamental principle. This was that space and time in both the worlds of appearance and reality were essentially Newtonian, thereby enabling simple *correlation* of the contents of the two worlds. From this perspective, the shift from Empiricism to Modified Empiricism is more drastic since it involves the reluctant abandonment of this principle of correlation. From 1919 onwards Russell was forced to accept that, at best, appearance could only be *partially correlated* with reality and, hence, that answering Kant would be much more difficult than previously imagined. As hinted in the previous chapter, Einstein's theories of space and time are largely responsible for this change.

Russell was enormously impressed by Einstein's work, acknowledging it as "one of the supreme triumphs of human genius" (*AMa*: 100). Responding in 1929 to a questionnaire from a Chicago magazine, Russell confessed that he "would gladly change places with about a dozen people – but first of all with Einstein" (Clark, 1975: 428).[1] (Russell warned the magazine that his responses were "truthful rather than interesting".) Earlier, in the summer of 1919, on being informed of the results of Eddington's eclipse observations which coincided with Einstein's predictions, Russell reportedly exclaimed: "To think that I have been spending all these years on muck!" (Wood, 1957: 133). Such reactions are, perhaps, typical both of Russell's tendency to disparage philosophy (including his own) and of his penchant for paying unbridled homage to the supreme intellectual achievements of others.

However, in the case of Einstein's work, there is much more to it than this. Russell had a long term preoccupation with developing close ties between philosophy and science. As we found above, he saw science as furnishing general results, which would be subject to philosophical analysis in order to arrive at more fundamental principles. One hazard of such a view is that, if the science with which philosophy starts undergoes drastic revision, then this will be liable to render obsolete the philosopher's previous efforts.[2] Russell came to realise that Einsteinian relativity entailed just such a drastic revision within science:

> A large part of the importance of relativity comes from the fact that it has substituted a single four-dimensional manifold (space-time) for the two manifolds, three-dimensional space and one-dimensional time. This is a change of structure, and therefore has far-reaching consequences. . . .
>
> (LA: 176).[3]

Obviously, then, we would expect Einsteinian relativity to have inspired major changes in Russell's philosophical views. Yet while freely acknowledging that "Einstein's revolution swept away everything at all resembling" the Kantian theory of geometry that he developed in the mid-1890's (*MPD*: 31), Russell was not quite as forthcoming on the influence of Einstein on his later realist views. Nevertheless this influence was profound. Without sweeping away everything, the effect of Einstein was to undermine some of the most central features of Russell's realism as it had evolved in the Empiricist phase. This effect is very clear when we recall that three main general principles underlie Russell's efforts to produce a major philosophical alternative to Kant. These three main principles are:

(i) a commitment to the general findings of science;
(ii) a conviction that private space and time can be closely related to their objective counterparts; and
(iii) a conviction that the nature of appearance enables us to infer something of the nature of its causal antecedents, particularly their relational characteristics.

Einsteinian relativity means that for the first time these three principles are in serious conflict with one another:

> . . . what we have hitherto regarded as the spatial and temporal properties of physical occurrences are found to be in large part dependent upon the observer; only a residue can be attributed to the occurrences in themselves, and only this residue can be involved in the formulation of any physical law which is to have an *a priori* chance of being true . . . Physics accounts both for the likenesses and for the differences between different people's perceptions of what we call the 'same' occurrence. But in order to do this it is first necessary for the physicist to find out just what are the likenesses. They are not quite those traditionally assumed, because neither space nor time separately can be taken as strictly objective. What is objective is a kind of mixture of the two called 'space-time'.
>
> (*ABCR*: 27).

(i) and (ii) are in conflict because, thanks to Einstein, there are no longer any simple objective counterparts to private space and time. The result is that they now have to be correlated with space-time and the correlation is not a close one. (iii) is also in trouble because the main relational characteristics of appearances are their spatial and temporal ones. Since these are now largely dependent on the observer, inferring something of the nature of the causal antecedents of the appearances becomes much more difficult. Clearly the programme of bridging appearance and reality via space and time needs drastic modification. How this is done will emerge from a detailed examination of how Russell applied each of the three principles (with necessary modifications) to his Modified Empiricist philosophising. Since space and time relations are no longer strictly objective, we will not be surprised to find them losing their status as exemplars of relations in general.

12.1 The Data from Science

It would be wrong to credit Einstein with all of the developments in science that shaped Russell's philosophy from the 1920's onwards. As we will find, quantum theory was also influential. Nevertheless it is Einstein's work that receives the overwhelming amount of attention in Russell's philosophical writings of the period. If the thesis that space and time theories are central to Russell's philosophising is correct, then Einstein having the lion's share of the attention is only to be expected.

What are the major outcomes of Einstein's work that Russell sees as crucial for philosophy? Firstly, as already indicated, "the most important novelty" is "the substitution of space-time for space and time" (PCR: 228).[4] Space and time intervals thereby lose their fundamental status in physics. Instead:

> Events have a four-dimensional order, by means of which we can say that an event *A* is nearer to an event *B* than to an event *C*; this is a purely ordinal matter, not involving anything quantitative. But, in addition, there is between neighbouring events a quantitative relation called 'interval', which fulfils the functions both of distance in space and of lapse of time in the traditional dynamics, but fulfils them with a difference.
>
> (PCR: 229).

Russell goes on to discuss time-like and space-like intervals. For events separated by a zero interval, there is no event between them in the four-dimensional space-time manifold. Thus, if "we hear a noise and see a colour simultaneously, our two perceptions are both in one place in space-time" (PCR: 229). For time-like intervals between two events, the time-order of the events is not ambiguous, but

> . . . whenever the interval between two events is space-like, their time-order will be different in different equally legitimate systems of measurement; in this case, therefore, the time-order does not represent a physical fact. . . . The fact that time is private to each body, not a single cosmic order, involves changes in the notions of substance and cause, and suggests the substitution of a series of events for a substance with changing states . . . Matter, also, may be reduced to a law according to which events succeed each other and spread out from centres. . . .
>
> (PCR: 229–30).[5]

Secondly, following Eddington, Russell maintains that a "surprising" outcome of relativity theory, which is "of great philosophical importance", is that

> . . . the traditional laws of physics, rightly understood, tell us almost nothing about the course of nature, being rather of the nature of logical truisms.
> (PCR: 230).

Although in his later works Russell retreated somewhat from calling physical laws "logical truisms", his conclusion, that "physics tells us much less about the real world than was formerly supposed" (PCR: 231), is a central theme for the Modified Empiricist phase.

A third important result of interest to philosophy is that relativity, with its successful account of gravitation, finally eliminates the concept 'force' from science:

> Bodies move as they do because that is the easiest possible movement in the region of space-time in which they find themselves, not because 'forces' operate upon them. The apparent need of forces to account for observed motions arises from mistaken insistence upon Euclidean geometry. . . .
> (PCR: 231).

A fourth important aspect of relativity theory, according to Russell, is that it adopts a *realist* view of the world.

> The 'observer' who is often mentioned in expositions of relativity need not be a mind, but may be a photographic plate or any kind of recording instrument. The fundamental assumption of relativity is realistic, namely, that those respects in which all observers agree when they record a given phenomenon may be regarded as objective, and not as contributed by the observers.
> (PCR: 231).

While common sense has always allowed for differences in, e.g. size and shape, arising from the particular point of view of the observer, relativity theory has demonstrated that

> . . . much more depends upon the point of view of the observer than was formerly thought. But there is found to be a residue which is not so dependent . . .
> (PCR: 232).

The implication here is that in his project of bridging appearance and reality, Russell in the Platonist and Empiricist phases had assumed too much of the data of experience to be objective. The task of the Modified Empiricist phase will be to identify the residue that is objective and, on that basis, to recommence the bridging.

Russell concludes that relativity theory leaves us with the following general picture of the world.

> In the four-dimensional space-time frame there are events everywhere, usually many events in a single place in space-time. The abstract mathematical

relations of these events proceed according to the laws of physics, but the intrinsic nature of the events is wholly and inevitably unknown except when they occur in a region where there is the sort of structure we call a brain. Then they become the familiar sights and sounds and so on of our daily life. We know what it is like to see a star, but we do not know the nature of the events which constitute the ray of light that travels from the star to our eye. And the space-time frame itself is known only in its abstract mathematical properties: there is no reason to suppose it similar in intrinsic character to the spatial and temporal relations of our perceptions as known in experience. There does not seem any possible way of overcoming this ignorance, since the very nature of physical reasoning allows only the most abstract inferences, and only the most abstract properties of our perceptions can be regarded as having objective validity.

(PCR: 232).

In the previous chapter, I argued that, contrary to those anxious to saddle the 1914 Russell with phenomenalism, he was committed in the Empiricist phase to the existence of *inferred* entities, i.e. entities that cannot be observed. As the last quotation suggests, by the Modified Empiricist phase, there are absolutely no doubts on this score. Russell has whole-heartedly embraced inferred events, with perceived events forming only a tiny sub-class of the manifold of events, i.e. the "world of physics contains more than the world of percepts" (*HK*: 340). Events are not simples, since Russell considered that there ". . . is a possibility of a further stage of analysis in which events are no longer the ultimate raw material" (*MPD*: 15).[6] However this further stage of analysis was, as far as I know, never seriously tackled. So that, for practical purposes, events are simples in the Modified Empiricist phase.

So far I have dwelt on the difficulties that Einsteinian relativity posed for Russell's project of bridging appearance and reality. However these developments in science also brought some advantages:

Relativity has introduced a wholly novel analysis of physical concepts, and has made it easier than it formerly was to build a bridge from physics to sense-data.

(*OKEW2*: 109).[7]

This is so because the

. . . substitution of space-time for space and time has made the category of substance less applicable than formerly, since the essence of substance was persistence through time, and there is now no one cosmic time. The result of this is to turn the physical world into a four-dimensional continuum of events, instead of a series of three-dimensional states of a world composed of persistent bits of matter.

(*OP*: 236).[8]

So, unlike in the Empiricist phase, where it was necessary to construct the basic substance-like particles of physics out of the non-substance-like sense-data and sensibilia of philosophy, in the Modified Empiricist phase, there is not such a disparity. The non-persistent events of relativity physics are strikingly similar to sense-data (or *percepts* as Russell usually terms them in the Modified Empiricist phase). So Russell, the arch-enemy of substance, is delighted that science has caught up with his notion of matter as a series of events and dispensed with particulars that persist through time.

12.2 Correlating Private Space and Time with Space-time

As stressed already, ". . . space . . . and time have ceased to be, for relativity physics, part of the bare bones of the world . . . " (*OKEW2*: 109), thereby seriously damaging the project of bridging appearance and reality by correlating their respective spaces and times. Whereas space and time survive in the world of appearance, in the world of reality ". . . space by itself and time by itself, are doomed to fade away into mere shadows and only a kind of union of the two will preserve an independent reality" (Minkowski, 1908: 75). Can private space and time be correlated with the new physical reality? Einsteinian relativity involves a possible infinity of both times and spaces in which moments and places can be located. Indeed relativity theory shows that the distinction between space and time is one that depends on the observer, e.g. what are seen as simultaneous, spatially separated events by observers in one reference frame, will appear as both spatially and temporally separate to observers in other reference frames. So the objectivity of both physical space and time, assumed by Russell in his Empiricist phase, was replaced by what he termed "physical subjectivity".[9] He uses this term to highlight the fact that the problem is *not* due to the subjectivity of the observer:

> It is natural to suppose that the observer is a human being, or at least a mind; but he is just as likely to be a photographic plate or a clock. That is to say, the odd results as to the difference between one 'point of view' and another are concerned with 'point of view' in a sense applicable to physical instruments just as much as to people with perceptions. The 'subjectivity' concerned in the theory of relativity is a *physical* subjectivity, which would exist equally if there were no such things as minds or senses in the world. (*ABCR*: 153).

In elucidating Russell's response to this physical subjectivity, it will be helpful to see what features of the Empiricist phase appearance/reality bridging remain. Firstly, the doctrine of the dual spaces and times is as strongly supported as ever and "failure to grasp the difference between them is a potent source of confusion" (*OP*: 109). However, as Table 9.1 indicates, in the Modified Empiricist phase, private space is now held to be absolute rather than relative. The space of the visual field is held to be, for "practical purposes", two-dimensional since visual far-and-near estimates are unreliable except over short distances.[10]

Secondly, the system of perspectives or points of view, with the inevitable comparison with Leibniz's monadology,[11] is retained. As before the perspectives are ordered according to differences of perspective, or causal laws, to constitute a three-dimensional physical space. It follows that the inferred perspective space of the Empiricist phase is just as *real* in the Modified Empiricist phase, though it is now viewed as a set of relations amongst groups of *events* rather than amongst sense-data and sensibilia.[12] As before, there are various ways of constructing points so as to convert this space to the absolute one of Newtonian physics.[13] Similar considerations apply to time and the construction of Newtonian instants (*HK*: 285–9). So, as before, private space and time can be correlated with physical counterparts:

> The theory of relativity does not affect the space and time of perception. My space and time, as known in perception, are correlated with those that, in physics, are appropriate to axes that move with my body. Relatively to axes tied to a given piece of matter, the old separation of space and time still holds; it is only when we compare two sets of axes in rapid relative motion that the problems arise which the theory of relativity solves. Since no two human beings have a relative velocity approaching that of light, comparison of their experiences will reveal no such discrepancies as would result if aeroplanes could move as fast as beta particles. In the psychological study of space and time, therefore, the theory of relativity may be ignored.
>
> (*HK*: 309).

So for observers in most terrestrial situations, where relativistic effects are negligible, the space and time of perception have an "objective counterpart" (*AMa*: 335) that correlates with them. (Once again, this supports the relational realism urged in Part One since the objective space and time are relative, i.e. they are nothing but relations between events.)

Does this constitute a satisfactory bridging of appearance and reality? Not for Russell. The problem is that appearance is here correlated with a small, atypical part of reality. He worries

> . . . whether any of the properties we assigned to the outer world were accepted without sufficient reason, merely because they were such as we think we find in the perceptual world. The subject is imaginatively difficult, and it is not easy to disentangle different levels of inference, but it is important to do so.
>
> (*AMa*: 333).

Russell was fearful that factors deriving from the characteristics of human beings would intrude into an account of the world, thereby distorting it. In his view, the human presence in the universe is insignificant:

> I accept without qualification the view that results from astronomy and geology, from which it would appear that there is no evidence of anything mental except in a tiny fragment of space-time, and that the great processes of nebular and stellar evolution proceed according to laws in which mind plays no part.
>
> (*MPD*: 12).

> Cosmically and causally, knowledge is an unimportant feature of the universe; a science which omitted to mention its occurrence might, from an impersonal point of view, suffer only from a very trivial imperfection.
>
> (*HK*: 9).

Russell's implied charge of "cosmic impiety" (*HWP*: 782) against John Dewey centres on the latter's philosophy being too preoccupied with human affairs. So there is no way Russell will be content with the limited results to be obtained from correlating perceptual space and time with objective space and time counterparts. He will be looking to connect the world of appearance with the large-scale structure of the universe of space-time as proclaimed by physics.

A different approach is suggested by relativity theory itself:

> Everything in relativity-theory goes (in a sense) from next to next; there are no direct relations between distant events, such as distance in time or space. And of course there are no forces acting at a distance . . . The interval between two neighbouring events is something objective, in the sense that any two careful observers will arrive at the same estimate of it. They will not arrive at the same estimate for the distance in space or the lapse of time between the two events, but the interval is a genuine physical fact, the same for all . . . But the interval is only definite when the two events are very near together; otherwise the interval depends upon the route chosen for travelling from the one event to the other.
>
> (*OP*: 88).[14]

So, according to Russell, relativity requires the "abandonment of all direct relations between distant events" (*AMa*: 55) because all such measures lack "physical significance":

> . . . relativity informs us that there is a residue of variability in measures which cannot be eliminated, because, in fact, the relations we try to measure are partially non-existent. Or, more correctly, they are relations involving more terms than we thought they did. We supposed that co-ordinates represented relations to the axes. But if we had two sets of axes momentarily coinciding, while one was moving relatively to the other, the co-ordinates of an event would not in general be the same with respect to both. And we cannot even, in any strict sense, discover any exact relation between distant points such as could give physical significance to coordinates. The appearance to the contrary is only an approximate truth, which cannot be made precise.
>
> (*AMa*: 337).

"All this", Russell concludes, "represents a failure of correspondence between physical space-time and perceptual space and time (*AMa*: 338).[15]

Clearly, if appearance and reality are to be bridged there will have to be a fresh approach to the problem. Since "the relations upon which space-time depends . . . [are] primarily confined to very small regions . . . " (*AMa*: 55), Russell abandons unrestricted space and time relations and shifts his attention to these more specific relations. How he does this will be best dealt with under the discussion of principle (iii), i.e. the one that concerns the causal antecedents of appearance. It is best left till then because relativity brings together geometry and physics (*ABCR*: 84–5), so that the small region relations upon which space-time depends are causal in nature. However before moving on to this, it is appropriate to say something about quantum theory. As already indicated, Russell expected quantum theory to have profound implications for philosophy. He saw it as supporting the four-dimensional manifold of events in space-time,[16] whilst also implying that physical phenomena are discontinuous. Accordingly, this led Russell to seek to ensure that his theory of space-time was compatible with both continuity and discontinuity.[17] All of this was done in the expectation that philosophy will require drastic changes in the future:

> The philosophy appropriate to quantum theory . . . has not yet been adequately developed. I suspect that it will demand even more radical departures from the traditional doctrine of space and time than those demanded by the theory of relativity.
>
> (*HWP*: 787).

12.3 Causal Laws and the Causal Antecedents of Appearances

As noted earlier in this chapter, Russell's principle (iii) (i.e. the view that the nature of appearance enables us to infer something of the nature of its causal antecedents, in particular their *relational characteristics*), is in some trouble from relativity theory. This is so because spatial and temporal characteristics are the ones involved. As we found in Chapter Nine, principle (iii) represents what Russell takes to be valid in the notion that primary qualities are objective. Now, however,

> Locke's half-way house has . . . been definitely abandoned . . . [because] . . . both space and time, as they occur in immediate experience, are recognised . . . as something quite different from the space-time which physics requires.
>
> (*AMa*: 132).

However, the same point represents even more trouble for principle (ii) (i.e. the view that the space and time of the two worlds closely correlate), so Russell has to rely on a modified form of (iii) to establish links between appearance and reality. The crucial notion is that of *causal law*, which Russell uses to establish the ordering of the four-dimensional manifold of events.

He defines *causal law* as follows:

> A causal law, as I use the term, is any law which, if true, makes it possible, given a certain number of events, to infer something about one or more other events.
>
> (*HK*: 344).[18]

Such laws assume an *ordering* of the events, i.e. each event has four coordinates.

> But the question arises: how do we assign co-ordinates to events in physical space-time? I maintain that, in doing so, we make use of causal laws. That is to say, the relation of causal laws to space-time order is a reciprocal one. The correct statement is: Events can be arranged in a four-dimensional order such that, when so arranged, they are interconnected by causal laws which are approximately continuous, i.e. events whose co-ordinates differ very little also differ very little. Or rather: Given any event, there is a series of closely similar events, in which the time-co-ordinate varies continuously from rather less to rather more than that of the given event, and in which the space-co-ordinates vary continuously about those of the given event. This principle, apparently, does not hold for quantum transitions, but it holds for macroscopic events, and for all events (such as light-waves) where there is no matter.
>
> (*HK*: 344).

So, according to Russell, there are two steps in locating events in physical space-time. The first "is only rough and approximate" (*HK*: 344) and involves assigning to events in physical space-time the same coordinates they have in perceptual space and time. This is said to be "very useful in establishing a first approximation" (*HK*: 345). The second step is to make corrections to the coordinates by means of *causal laws*. In determining causal laws, Russell seeks an empirical relation that will fit the requirements of relativistic physics, e.g. it should be a relation that holds between closely neighbouring events but not between distant events (i.e. unlike space and time relations). Such an empirical relation is compresence. Compresent events "overlap in space-time", or, in experiential terms:

> Two events are 'compresent' when they are related in the way in which two simultaneous parts of one experience are related. At any given moment, I am seeing certain things, hearing others, touching others, remembering others, and expecting yet others. All these percepts, recollections, and expectations are happening to me now; I shall say that they are mutually 'compresent'. I assume that this relation, which I know in my own experience, can also hold between events that are not experienced, and can be the relation by which space-time order is constructed.
>
> (*HK*: 347).

Russell goes on to argue that the relation of compresence, supplemented by suitable axioms, is sufficient to generate space-time order, an order that Einstein's relation of *interval* cannot generate.

Where does this leave us in the general project of bridging appearance and reality, a project that I have argued to be Russell's major ongoing philosophical concern?

> According to the above theory, there are certain elements that are carried over unchanged from the world of sense to the world of physics. These are: the relation of compresence, the relation of earlier and later, some elements of structure, and differences in certain circumstances – i.e. when we experience different sensations belonging to the same sense, we may assume that their causes differ. This is the residue of naive realism that survives in physics.
>
> (*HK*: 349).

When we recall that prior to the Modified Empiricist phase extended spatial and temporal relations were the major components of such a list, the impact of principle (ii) being largely abandoned and principle (iii) being severely modified is clear. This is the basis of Russell's oft-repeated claim in the Modified Empiricist phase that physics provides us with extremely limited knowledge about the world:

> What physics says about the world is much more abstract than it seems to be, because we imagine that its space is what we know in our own experience, and that its matter is the kind of thing that feels hard when we touch it. In fact, even assuming physics true, what we know about the physical world is very little.
>
> (*HK*: 345).[19]

There is, of course, a lot more that could be said about causal laws,[20] but I take it that enough has been done to demonstrate that they replace space and time relations as the means of bridging appearance and reality. As a result, the correlation between the two turns out to be a lot more tenuous than what Russell had optimistically expected to achieve when he first set out to answer Kant. The dethroning of space and time relations as *exemplars* of relations is complete.

12.4 The Influence of Relativity Theory on Russell's Wider Philosophy

In the last chapter we found Russell's tendency to view all relations as space-like and time-like starting to wane. With space and time relations no longer enjoying a special status amongst relations in the Modified Empiricist phase, this trend accelerates. Thus *acquaintance*, the model of the basic dyadic relation in the Empiricist phase, becomes *noticing* when Russell adopts neutral monism. While there is still a duality involved in noticing, it is now "a matter of degree" which cannot be analysed as a simple dyadic relation. In fact Russell takes noticing to be an undefined term.[21] Similarly *belief* and *memory* are provided with new accounts. When Russell earlier viewed all relations as space-like and time-like, we found him construing belief (judgment) and memory as direct relations between egos and objects in the world, whether present or past. As Pears put it, his theory of mind aimed "to achieve maximum exposure to the external world" (1967: 198). We found the various versions of the multiple relation theory of judgment to be failed attempts to retain this principle. Now, with the abandonment of the ego, belief, memory and the like are no longer viewed as dyadic relations based on acquaintance. From now on *images* play an essential role in Russell's philosophy of mind and the parallels with his space and time theories disappear.[22] Likewise, with space and time relations under a cloud, Russell is even more wary of making metaphysical claims about *universals* in general. Hence his preference, discussed in Chapter Eight, for confining his attention in the Modified Empiricist phase to *relation-words* rather than *relations*, and "technical" or "logical" questions, rather than "metaphysical" ones. In all of this, the fall from grace of space and time relations is evident.

Throughout Part Two, it has been argued that Russell's efforts to bridge appearance and reality have significantly influenced wider areas of his philosophy. We have seen a series of examples of this from the years when space and time relations were the focal point of such bridging. Now that causal laws have displaced space and time relations as the basis of appearance-reality bridging, it comes as no surprise to find that this change is responsible for some of the more novel doctrines that Russell puts forward in the Modified Empiricist phase. A prime example is his notorious "under the hat" theory that what you see is in your brain:

> . . . what the physiologist sees when he looks at a brain is part of his own brain, not part of the brain he is examining.
>
> (*AMa*: 383).

This theory appears to have attracted almost universal critical condemnation,[23] yet the critics are generally unaware of how it hinges on what Russell takes to be the philosophical implications of space-time.[24] Russell couldn't be more explicit on this, as the continuation of the previous quotation shows:

> The feeling of paradox about this view comes, I should say, from wrong views of space. It is true that what we see is not located where our percept of our own brain would be located if we could see our own brain; but this is a question of perceptual space, not of the space of physics. The space of physics is connected with causation in a manner which compels us to hold that our percepts are in our brains, if we accept the causal theory of perception, as I think we are bound to do. To say that two events have no spatio-temporal separation is to say that they are compresent; to say that they have a small separation is to say they are connected by causal chains all of which are short. The percept must therefore be nearer to the sense-organ than to the physical object, nearer to the nerve than to the sense-organ, and nearer to the cerebral end of the nerve than to the other end. This is inevitable, unless we are going to say that the percept is not in space-time at all.
>
> (*AMa*: 383).[25]

But, of course, Russell has repeatedly argued that percepts *are* in space-time. If a visual field is a point-instant in physical space-time, as Russell maintains, and if the spatial and temporal relations of the visual field don't correspond to relations in physical space-time, then the explanation of perception as an instance of causal laws starts to look less idiosyncratic.

Similar considerations apply to the puzzling Modified Empiricist phase doctrine that qualities, such as whiteness, are particulars.[26] The crucial point, as Russell stresses in his 1955 addendum to "On the Relations of Universals and Particulars", is that he no longer thinks "that there are any spatial or temporal relations which always and necessarily imply diversity". Prior to the Modified Empiricist phase, spatially separated instances of whiteness in the visual field were taken to reflect spatial separation of their causes in physical space. Now, however, spatial relations in the visual field have no close counterpart in space-time. According to Russell the momentary visual field is a complete complex of compresence, i.e. a space-time point-instant,[27] while an instance of (say) whiteness in the visual field is a complex of compresence (incomplete).[28] So separate instances of whiteness in the same visual field belong to the one complete complex of compresence, i.e. the one space-time point-instant. Once this is understood it is, perhaps, not so far-fetched to suggest that the correlates of qualities might be particulars in the four-dimensional event manifold of relativity physics.

None of this amounts to a *defence* of the "under the hat" theory or the "qualities as particulars" theory. That would take us beyond our present concerns. What is evident, however, is that once again Russell's space-time theories significantly shape his wider philosophy.

I will conclude with a comment on why I date the Modified Empiricist phase from 1919, since Maxwell has claimed that 1924 is the significant turning point (1974: 170). If it is true, as argued at length in the previous chapter that Russell only dispensed with the ego when he realised that the inferred perspective space has no close correlation with space-time of Einsteinian physics, then this must have occurred about 1919. This is so because in the 1919 paper, "On Propositions: what they are and how they mean", Russell dispenses with the ego in favour of neutral monism, uses the notion of causal laws (OPWH: 289ff.)[29] and abandons the multiple relation theory of belief in favour of one involving images (OPWH: 295–6). Although relativity is not mentioned in this article, Russell shortly after published one discussing relativity.[30] *Analysis of Mind* (published in 1921) gives explicit comment on the space-time views characteristic of the Modified Empiricist phase,[31] thus further supporting my 1919 dating.

NOTES

1. See also PAP: 126.
2. Russell was not too worried by this. Cf. LA: 176: "Philosophy . . . should be bold in suggesting hypotheses as to the universe which science is not yet in a position to confirm or confute. But these should always be presented as hypotheses, not (as is too often done) as immutable certainties like the dogmas of religion." See also *AMa*: 394.
3. See also *OP*: 86.
4. PCR (which first appeared in *Encyclopedia Britannica*, 13th Edition, 1926, as "Relativity: Philosophical Consequences"), is a very useful short statement by Russell of the main results of relativity that have significance for his philosophy.
5. Russell here seems to have adopted a new account of what it is to be a physical fact. The

point that he makes here applied equally to sens-data in 1914 (with 'perspectives' replacing 'legitimate systems of measurement') yet sens-data remained physical. (I owe this point to Nicolas Griffin.)

6. See also *AMa*: 277ff. and 293.
7. It is instructive to compare the paragraph from which this passage comes (*OKEW2*: 108–109) with the one it replaces (*OKEW1*: 103–4).
8. See also *OP*: 125 and *AMa*: 244.
9. MPP: 302; *ABCR*: 153 and *AMa*: 224. See also *OP*: 87 and *HWP*: 786.
10. *AMa*: 166 and *HK*: 234. See also *OP*: 109. As far as I can tell, Russell never entertained the idea that the visual field might be two-dimensional in his earlier phases. This seems strange, given that there has been, after all, a major tradition in the phenomenology of vision that considers the visual field to be two-dimensional with the third dimension being an intellectual construction not given by sight.
11. RTC: 708–9 and *MPD*: 17–8.
12. See, e.g., *HK*: Part III, Ch. VI and Part IV, Ch. VI.
13. See, e.g., *HK*: 295–8.
14. See also *ABCR*: 155 and PAP: 127.
15. The context makes it absolutely clear that 'perceptual' is intended rather than 'perpetual' (sic). In *HK*: 345, Russell says that the correlation of the two is "only rough . . . " with ". . . subsequent corrections . . . effected by means of causal laws", i.e. new sorts of relations are needed to bridge appearance and reality, as we will find shortly.
16. RTC: 701 and *HWP*: 786.
17. See, e.g., *AMa*: 292, 311 and 341.
18. See also *AMa*: Ch. XXX; RTC: 705ff; *MPD*: Ch. 2.
19. See also *ABCR*: 157–8; RTC: 706; *MPD*: 13; and *Auto.*: 395.
20. Russell says a lot more, e.g., in *HK*: Part VI. See also Darlington (1978).
21. A good brief discussion of these developments is *MPD*: 103–7.
22. Detailed discussion occurs in Pears, 1974: 117–38.
23. See, e.g., O'Connor (1979).
24. Cf. Russell's 1944 claim: "I have not so far found any philosopher who knew what I meant by this statement" (RTC: 705).
25. See also *OP*: 110–2; RTC: 705–6; and *MPD*: 18–9 for explicit statements of the same points.
26. See, e.g., *HK*: Part IV, Ch. VIII; and *MPD*: Ch. 14.
27. See, e.g., *HK*: 313.
28. See, e.g., *HK*: 316.
29. Russell uses causal laws to relate words to their *meanings*. Incidentally, I believe that a study of Russell's changing views on meaning would further support the main conclusions of this book.
30. "Relativity Theory of Gravitation", *English Review*, Vol. 30, Jan. 1920: 11–8.
31. See, e.g., Preface and p. 127.

CHAPTER 13

Conclusion

This book has examined the general outlines of Russell's philosophy from two different perspectives. In Part One, an elucidation of Russellian analysis was presented. It was shown that, partly stemming from work in his neo-Hegelian phase, Russell evolved a distinctive method of philosophical analysis that grounded all of his subsequent philosophy following the 1900 revolt into realism. This account of analysis explained the prominence of relations in the various facets of his work. Despite this, it was argued, relations have not generally received the attention they deserve from Russell's commentators. Several instances of this omission leading to serious misinterpretations of his philosophical position were highlighted. The theory of logical constructions, for instance, was shown to be one central component of Russell's philosophy in which relations play a crucial part. Yet, as was also shown, failure to recognise the role of relations in logical constructions has been the fundamental cause of rife misunderstandings of this central aspect of his work.

In Part Two, it was argued that underlying the *continuity* of the method of analysis, and its tendency to proliferate relations, was a series of highly significant *changes*. These changes were the result of Russell being confronted with irresistible arguments for altering his views on space and time. The wide-ranging impact of these changes on his other views has been outlined.[1] The centrality of space and time doctrines to Russell's philosophy was traced to his continuing efforts to provide an alternative to Kant. This also accounted for space and time relations starting out as exemplars of relations, but then losing their importance as the task of answering Kant became increasingly more complex than Russell had anticipated. Just as later developments in mathematics and science posed difficulties for the classical Kantian doctrines, so also developments in science, such as those due to Einstein, rendered obsolete the simple outlines of Russell's proposal for answering Kant. His final position is thereby much more complex and much less elegant than the one that he had hoped for when he first revolted into realism.

The perspectives on Russell's philosophy set out in the two parts of this book obviously reinforce and complement one another. Each of the three phases, dealt with in detail in Part Two, respectively imply that different combinations of basic entities and relations will be central to examples of *analysis* carried out in any particular phase. Thus, while the structure of Russellian analysis, as elucidated in Part One, remains constant, the changes explained in Part Two account for the major differences between actual philosophical analyses undertaken in different phases.

NOTE

1. Hager (1987) provides an example of how clarity about Russell's phases and the differences between them can remove confusion in other areas.

Bibliography

I. Works by Russell

Abbreviations for Titles of Russell's Works

The following is a key to abbreviations for books and papers by Russell as used for citation in the text. One problem that strikes anyone familiar with the literature on Russell's work is that the pagination frequently differs between various editions or publications of a particular work. Accordingly, besides the abbreviations, the following list shows, firstly, the publication details of the edition or printing that *I have used* and, where this is not the first edition, the date of *original* publication.

1. *The Collected Papers of Bertrand Russell.*

Papers 1	Vol. 1 *Cambridge Essays: 1888–99*, Kenneth Blackwell *et al.* (eds.) (London: Allen & Unwin, 1983)
Papers 2	Vol. 2 *Philosophical Papers: 1896–99*, Nicholas Griffin and Albert C. Lewis (eds.) (London: Unwin Hyman, 1990).
Papers 3	Vol. 3. *Towards the "Principles of Mathematics": 1900–02*, Gregory H. Moore (ed.) (London: Routledge, 1993).
Papers 4	Vol. 4. *Foundations of Logic: 1903–05*, Alasdair Urquhart (ed.) (London: Routledge, 1993).
Papers 6	Vol. 6. *Logical and Philosophical Papers: 1909–13*, John Slater (ed.) (London: Routledge, 1992).
Papers 7	Vol. 7. *Theory of Knowledge: The 1913 Manuscript*, Elizabeth Eames and Kenneth Blackwell (eds.) (London: Allen & Unwin, 1986).
Papers 8	Vol. 8. *The Philosophy of Logical Atomism and Other Essays: 1914–19*, John Slater (ed.) (London: Allen & Unwin, 1986).
Papers 9	Vol. 9. *Essays on Language, Mind and Matter: 1916–26*, John Slater (ed.) (London: Unwin Hyman, 1988).
Papers 12	Vol. 12. *Contemplation and Action: 1902–14*, Richard Rempel *et al.* (eds.) (London: Allen & Unwin, 1985).
Papers 13	Vol. 13. *Prophecy and Dissent: 1914–16*, Richard Rempel *et al.* (eds.) (London: Unwin Hyman, 1988).

2. *Other Works of Bertrand Russell.*

ABCR	*ABC of Relativity*, (London: Allen & Unwin, 1977; 1st edn., 1925).
AMa	*The Analysis of Matter*, (London: Allen & Unwin-Dover, 1954; 1st edn., 1927).

AMi — *The Analysis of Mind*, (London and New York: Allen & Unwin/Humanities Press, 1971; 1st edn., 1921).

AMR — "An Analysis of Mathematical Reasoning", 1898, *Papers* 2, 163–222.

AMR/TS — "An Analysis of Mathematical Reasoning", TS, 1898, *Papers* 2, 223–38.

APG — "The *A Priori* in Geometry", 1896, *Papers* 1, 291–304.

AR — "Analytic Realism", 1911, *Papers* 6, 132–46.

Auto. — *The Autobiography of Bertrand Russell*, 1 vol., (London: Allen & Unwin, 1975; 1st edn., 1967, 1968, 1969).

DMP — "Definitions and Methodological Principles in Theory of Knowledge", 1914, *Monist*, 24, 582–93.

EA — *Essays in Analysis*, D. Lackey (ed.) (London: Allen & Unwin, 1973).

EAE — "Are Euclid's Axioms Empirical?", 1898, *Papers* 2, 325–38.

EFG — *An Essay on the Foundations of Geometry* (Cambridge: CUP, 1897).

ERB — "Some Explanations in Reply to Mr Bradley", 1910, *Papers* 6, 354–58.

FIAM — "The Fundamental Ideas and Axioms of Mathematics", 1899, *Papers* 2, 265–305.

HK — *Human Knowledge: Its Scope and Limits* (London and New York: Allen & Unwin/ Humanities Press, 1966; 1st edn., 1948).

HWP — *History of Western Philosophy* (London: Allen & Unwin, 1971; 1st edn., 1946).

IMP — *An Introduction to Mathematical Philosophy* (London: Allen & Unwin, 1970; 1st edn., 1919).

IMT — *An Inquiry into Meaning and Truth* (London: Allen & Unwin, 1966; 1st edn., 1940).

KAKD — "Knowledge by Acquaintance and Knowledge by Description", 1911, *Papers* 6, 148–61.

LA — "Logical Atomism", 1924, *Papers* 9, 162–79.

LE — "The Limits of Empiricism", 1936, *Proceedings of the Aristotelian Society*, N.S., XXXVI, 131–50.

LK — *Logic and Knowledge: Essays 1901–1950*, R.C. Marsh (ed.) (London: Allen & Unwin, 1956).

LOG — "The Logic of Geometry", 1896, *Papers* 1, 267–86.

LOR — "The Logic of Relations with Some Applications to the Theory of Series", 1901, *Papers* 3, 314–49.

LP — "Logical Positivism", 1950 in *LK*, 367–82.

ML — *Mysticism and Logic and Other Essays* (London: Allen & Unwin, 1969; 1st edn., 1917).

MMD — "My Mental Development", 1944 in Schilpp (ed.), 1944, 1–20.

MOP — *My Own Philosophy* (Hamilton, Ontario: McMaster University Library Press, 1972; written in 1946).

MPD — *My Philosophical Development* (London: Allen & Unwin, 1975, 1st edn., 1959).

MPL — "Is Mathematics Purely Linguistic?", 1950–1952 in *EA*, 295–306.

MPP — "Materialism, Past and Present", 1924, *Papers* 9, 294–302.

MTCA — "Meinong's Theory of Complexes and Assumptions", 1904 in *EA*, 21–76.

MTT — "The Monistic Theory of Truth", 1906 in *PE*, 131–46.

NSD — "The Nature of Sense-Data: A Reply to Dr Dawes Hicks", 1913, *Papers* 6, 184–89.

NTF — "On the Nature of Truth and Falsehood", 1910, *Papers* 6, 116–24.

OD — "On Denoting", 1905, *Papers* 4, 415–27.

OET — "On the Experience of Time", 1915, *Monist*, 25, 212–33.

OKEW — *Our Knowledge of the External World* (used where the edition doesn't matter).

OKEW1 — *Our Knowledge of the External World as a Field for Scientific Method in Philosophy* (Chicago and London: The Open Court Publishing Co., 1914).

OKEW2 — *Our Knowledge of the External World as a Field for Scientific Method in Philosophy* (London: Allen & Unwin, revised edn., 1926).

OKEW3 — *Our Knowledge of the External World as a Field for Scientific Method in Philosophy* (New York: Norton & Co., revised edn., 1929).

ONA	"On the Nature of Acquaintance", 1914 in *LK*, 127–74.
ONO	"On the Notion of Order", 1901, *Papers* 3, 291–309.
OP	*An Outline of Philosophy* (London: Allen & Unwin, 1979; 1st edn., 1927).
OPWH	"On Propositions: What They Are and How They Mean", 1919, *Papers* 8, 276–306.
OQ	"On Quantity and Allied Conceptions", 1898, *Papers* 2, 117–35.
PAP	"Physics and Perception", 1922, *Papers* 9, 126–33.
PCR	"Philosophical Consequences of Relativity", 1926, *Papers* 9, 228–32.
PE	*Philosophical Essays* (London: Allen & Unwin, revised edition, 1966; 1st edn., London: Longmans, Green & Co., 1910).
PFL	"Philosophy for Laymen", 1950 in *UE*, 27–37.
PfM	*Portraits from Memory and Other Essays* (London: Allen & Unwin, 1957; 1st edn., 1956).
PIML	"The Philosophical Importance of Mathematical Logic", 1911, *Papers* 6, 33–40.
PL	*Critical Exposition of the Philosophy of Leibniz* (London: Allen & Unwin, 1975; 1st edn., 1900).
PLA	"The Philosophy of Logical Atomism", 1918, *Papers* 8, 160–244.
PM	*Principia Mathematica*, with A.N. Whitehead, 3 vols. (Cambridge: CUP, 1925–7; 1st edn., 1910–13).
POB	"The Philosophy of Bergson", 1912, *Papers* 6, 320–37.
PoM	*The Principles of Mathematics* (New York: Norton, 1964, 1st edn., 1903).
PP	*The Problems of Philosophy* (London: Oxford University Press, 1962; 1st edn., 1912).
PTC	"Philosophy in the Twentieth Century", 1924, *Papers* 9, 451–66.
PTS	"Is Position in Time and Space Absolute or Relative?", 1901, *Papers* 3, 261–82.
PU	"The Problem of Universals", 1946, *Polemic*, No.2, 21–35.
PUM	"Philosophy's Ulterior Motives", 1937 in *UE*, 48–58.
RMDP	"The Regressive Method of Discovering the Premises of Mathematics", 1911 in *EA*, 272–83.
RMU	"Review of: A. Meinong, Untersuchungen zur Gegenstandstheorie und Psychologie", 1905 in *EA*, 77–88.
RNQ	"On the Relations of Number and Quantity", 1897, *Papers* 2, 70–82.
RSDP	"The Relation of Sense-Data to Physics", 1914, *Papers* 8, 5–26.
RTC	"Reply to Criticisms", 1944 in Schilpp (ed.), 1944, 680–741.
RUP	"On the Relations of Universals and Particulars", 1912, *Papers* 6, 167–82.
SAI	"Sensation and Imagination", 1915, *Monist*, 25, 28–44.
SE	*Sceptical Essays* (London: Allen & Unwin, 1970; 1st edn., 1928).
SH	"Science and Hypothesis (Review of Poincaré)", 1905, *Papers* 4, 589–94.
SMP	"On Scientific Method in Philosophy", 1914, *Papers* 8, 57–73.
SOM	"The Study of Mathematics", 1907, *Papers* 12, 85–93.
UCM	"The Ultimate Constituents of Matter", 1915, *Papers* 8, 75–86.
UE	*Unpopular Essays* (London: Allen & Unwin, 1970; 1st edn., 1950).

II. Works by Other Authors

Aristotle (1928) *Metaphysica* (trans. W.D. Ross, 2nd. edn.) (Oxford: Clarendon Press).

Armstrong D.M. (1978) *Universals and Scientific Realism* (Vol. 1, Nominalism and Realism; Vol. 2, A Theory of Universals) (Cambridge: Cambridge University Press).

Ayer A.J. (1971) *Russell and Moore, The Analytic Heritage* (London: Macmillan).

Ayer A.J. (1972) *Russell* (London: Collins).

Black M. (1971) "The Elusiveness of Sets", *Review of Metaphysics*, 24, 614–36.

Broad C.D. (1957) "The Local Historical Background of Contemporary Cambridge Philosophy" in Mace (ed.), 13–61.

Brown S. (1978) *Realism and Logical Analysis* (Buckinghamshire: Open University Press).

Cassin C.E. (1970) "Russell's Discussion of Meaning and Denotation: a Re-examination" in Klemke (ed.), 256–72.

Clack R. (1969) *Bertrand Russell's Philosophy of Language* (The Hague: Nijhoff).

Clark R.W. (1975) *The Life of Bertrand Russell* (London: Jonathan Cape and Weidenfeld and Nicholson).

Cocchiarella N.B. (1980) "The Development of the Theory of Logical Types and the Notion of a Logical Subject in Russell's Early Philosophy", *Synthese*, 45, 71–115. (In Cocchiarella (1987) as Chapter 1.)

Cocchiarella N.B. (1982) "Meinong Reconstructed Versus Early Russell Reconstructed", *Journal of Philosophical Logic*, 11, 183–214. (In Cocchiarella (1987) as Chapter 3.)

Cocchiarella N.B. (1986) Logical Investigations in Predication Theory and the Problem of Universals (Naples: Bibliopolis).

Cocchiarella N.B. (1987) *Logical Studies in Early Analytic Philosophy* (Columbus, Ohio: Ohio State University Press). (Esp. Chapter 2, pp. 64–118, titled 'Frege, Russell, and Logicism: A Logical Reconstruction' and Chapter 5, pp. 193–221, titled 'Russell's Theory of Logical Types and the Atomistic Hierarchy of Sentences') (Chapter 5 is also in Savage and Anderson (eds.).)

Copleston F. (1964) *A History of Philosophy*, Vol. 6, Part II (New York: Doubleday).

Copleston F. (1967) *A History of Philosophy*, Vol. 8, Part II (New York: Doubleday).

Darlington E.W. (1978) "Bertrand Russell's Causal Theories", M.A. Thesis, McMaster University.

Demopoulos W. (1981) "New Work on Russell's Early Philosophy", *Russell: The Journal of the Bertrand Russell Archives*, N.S. 1(2), 163–70.

Dorward A. (1951) *Bertrand Russell* (London: Longman's, Green & Co.).

Eames E.R. (1969) *Bertrand Russell's Theory of Knowledge* (London: Allen & Unwin).

Eames E.R. (1984) Editorial Introduction to *Papers* 7.

Eames E.R. (1989) *Bertrand Russell's Dialogue with His Contemporaries* (Carbondale and Edwardsville: Southern Illinois University Press).

Feinberg B. and Kasrils R. (1973) *Bertrand Russell's America, Vol. 1: 1896–1945* (London: Allen & Unwin).

Flew A. (1979) *A Dictionary of Philosophy* (London: Pan Books).

Foster T.R. (1983) "Russell on Particularized Relations", *Russell: The Journal of the Bertrand Russell Archives*, N.S. 3(2), 129–43.

Fritz C.A. (1952) *Bertrand Russell's Construction of the External World* (London: Routledge & Kegan Paul).

Goldfarb W. (1989) "Russell's Reasons for Ramification" in Savage and Anderson (eds.), 24–40.

Grattan-Guinness I. (1974) "The Russell Archives; Some New Light on Russell's Logicism", *Ann. Sci.*, 31, 387–406.

Grattan-Guinness I. (1976) "On the Mathematical and Philosophical Background to Russell's *The Principles of Mathematics*" in Thomas and Blackwell (eds.), 157–73.

Grattan-Guinness I. (1977) *Dear Russell – Dear Jourdain* (London: Duckworth).

Grattan-Guinness I. (1980) "Georg Cantor's Influence on Bertrand Russell", *History and Philosophy of Logic*, 1, 61–93.

Grattan-Guinness I. (1984) "Notes on the Fate of Logicism from *Principia Mathematica* to Godel's Incompletability Theorem", *History and Philosophy of Logic*, 5, 67–78.

Grattan-Guinness I. (1990) "Bertrand Russell (1872–1970) After Twenty Years", *Notes Rec. R. Soc. Lond.*, 44, 280–306.

Griffin N. (1980) "Russell on the Nature of Logic (1903–1913)", *Synthese*, 45, 117–88.

Griffin N. (1982) "New Work on Russell's Early Philosophy", *Russell: The Journal of the Bertrand Russell Archives*, N.S. 2(2), 69–83.

Griffin N. (1983) "Reply to Demopoulos", *Russell: The Journal of the Bertrand Russell Archives*, N.S. 3(1), 43–7.

Griffin N. (1985) "Russell's Multiple Relation Theory of Judgment", *Philosophical Studies*, 47, 213–47.

Griffin N. (1986) "Wittgenstein's Criticism of Russell's Theory of Judgment", *Russell: The Journal of the Bertrand Russell Archives*, N.S. 5(2), 132–45.

Griffin N. (1991) *Russell's Idealist Apprenticeship* (Oxford: Clarendon Press).

Griffin N. (ed.) (1992) *The Selected Letters of Bertrand Russell* (Volume 1) (Boston/New York: Houghton Mifflin).

Griffin N. (1992) "The Legacy of Russell's Idealism for His Later Philosophy: The Problem of Substance", *Russell: The Journal of the Bertrand Russell Archives*, N.S. 12(2), 186–96.

Griffin N. (1993a) "Terms, Relations, Complexes" in Irvine and Wedeking (eds.), 159–92.

Griffin N. (1993b) "Denoting Concepts in *The Principles of Mathematics*", unpublished manuscript.

Hager P.J. (1979) "Russell Resurrected", *Proceedings of the Russellian Society* (University of Sydney), 4, 1–13.

Hager P.J. (1987) "Russell and Zeno's Arrow Paradox", *Russell: The Journal of the Bertrand Russell Archives*, N.S. 7(1), 3–10.

Hamlyn D.W. (1970) *The Theory of Knowledge* (London: Macmillan).

Harman G. (1965) "The Inference to the Best Explanation", *Philosophical Review*, 74(1), 88–95.

Hintikka J. (ed.) (1980) *Bertrand Russell's Early Philosophy* (Part I), *Synthese*, 45(1).

Hintikka J. (ed.) (1981) *Bertrand Russell's Early Philosophy* (Part II), *Synthese*, 46(2).

Honderich T. (ed.) (1984) *Philosophy Through Its Past* (Harmondsworth: Penguin).

Hylton P. (1990) *Russell, Idealism and the Emergence of Analytic Philosophy* (Oxford: Clarendon Press).

Irvine A.D. (1989) "Epistemic Logicism & Russell's Regressive Method", *Philosophical Studies*, 55, 303–27.

Irvine A.D. and Wedeking G.A. (eds.) (1993) *Russell and Analytic Philosophy* (Toronto: University of Toronto Press).

Jager R. (1972) *The Development of Bertrand Russell's Philosophy* (London: Allen & Unwin).

Kilmister C.W. (1984) *Russell* (Philosophers in Context series) (London: Harvester).

Klemke E.D. (ed.) (1970) *Essays on Bertrand Russell* (Chicago: University of Illinois).

Kuntz P.G. (1986) *Bertrand Russell* (Boston: Twayne).

Lakatos I. (Worrall J. and Zahar E. (eds.)) (1976) *Proofs and Refutations* (Cambridge: Cambridge University Press).

Lakatos I. (Worrall J. and Currie G. (eds.)) (1978a) *The Methodology of Scientific Research Programmes* (Philosophical Papers Vol. 1) (Cambridge: Cambridge University Press).

Lakatos I. (Worrall J. and Currie G. (eds.)) (1978b) *Mathematics, Science and Epistemology* (Philosophical Papers Vol. 2) (Cambridge: Cambridge University Press).

Landini G. (1987) "Russell's Substitutional Theory of Classes and Relations", *History and Philosophy of Logic*, 8, 171–200.

Landini G. (1989) "New Evidence Concerning Russell's Substitutional Theory of Classes", *Russell: The Journal of the Bertrand Russell Archives*, N.S. 9(1), 26–42.

Landini G. (1991) "A New Interpretation of Russell's Multiple-Relation Theory of Judgment", *History and Philosophy of Logic*, 12, 37–69.

Lewis J. (1968) *Bertrand Russell: Philosopher and Humanist* (London: Lawrence and Wishart).

Lowe V. (1974) "Whitehead's 1911 Criticism of *The Problems of Philosophy*", *Russell: The Journal of the Bertrand Russell Archives*, 13, 3–10.

Lycan W.G. (1981) "Logical Atomism and Ontological Atoms", *Synthese*, 46, 207–29.

Mace C.A. (ed.) (1957) *British Philosophy in the Mid-Century* (London: Allen & Unwin).

Magee B. (1971) *Modern British Philosophy* (London: Secker & Warburg).

Magee B. (1982) *Men of Ideas* (Oxford: Oxford University Press).

Maxwell G. (1974) "The Later Bertrand Russell: Philosophical Revolutionary" in Nakhnikian (ed.), 169–82.

McHenry L.B. (1992) *Whitehead and Bradley: A Comparative Analysis* (Albany: SUNY Press).

Minkowski H. (1908) "Space and Time" in H.A. Lorentz *et al.* (1952), *The Principle of Relativity* (New York: Dover), 73–91.

Moore G.E. (1899) "The Nature of Judgement", *Mind*, N.S. 8, 176–93.

Moore G.H. (1988) "The Roots of Russell's Paradox", *Russell: The Journal of the Bertrand Russell Archives*, N.S. 8(1–2), 46–56.

Muirhead J.H. (ed.) (1924) *Contemporary British Philosophy* (First Series) (London: Allen & Unwin).

Nakhnikian G. (ed.) (1974) *Bertrand Russell's Philosophy* (London: Duckworth).

Nusenoff R.E. (1978) "Russell's External World: 1912–1921", *Russell: The Journal of The Bertrand Russell Archives*, 29–32, 65–82.

O'Connor D.J. (1979) "Russell's Theory of Perception" in Roberts (ed.), 304–20.

Passmore J. (1969) "Russell and Bradley" in R. Brown and C.D. Rollins (eds.), *Contemporary Philosophy in Australia* (London: Allen & Unwin), 21–30.

Pears D.F. (1967) *Bertrand Russell and the British Tradition in Philosophy* (London: Collins).

Pears D.F. (ed.) (1972) *Bertrand Russell: A Collection of Critical Essays* (New York: Doubleday).

Pears D.F. (1974) "Russell's Theories of Memory 1912–1921" in Nakhnikian (ed.), 117–38

Popper K. (1968) *The Logic of Scientific Discovery* (London: Hutchinson).

Quine W.V. (1972) "Russell's Ontological Development" in Pears (ed.), 290–304.

Roberts G.W. (ed.) (1979) *Bertrand Russell Memorial Volume* (London: Allen & Unwin).

Rodriguez-Consuegra F. (1987) "Russell's Logicist Definitions of Numbers: Chronology and Significance", *History and Philosophy of Logic*, 8, 141–69.

Rodriguez-Consuegra F. (1989a) "Russell's Theory of Types, 1901–1910: Its Complex Origins in the Unpublished Manuscripts", *History and Philosophy of Logic*, 10, 131–64.

Rodriguez-Consuegra F. (1989b) "The Origins of Russell's Theory of Descriptions According to the Unpublished Manuscripts", *Russell: The Journal of the Bertrand Russell Archives*, N.S. 9(2), 99–132.

Rodriguez-Consuegra F. (1991) *The Mathematical Philosophy of Bertrand Russell: Origins and Development* (Basel, Boston and Berlin: Birkhauser).

Rodriguez-Consuegra F. (1992) "A New Angle on Russell's 'Inextricable Tangle' over Meaning and Denotation", *Russell: The Journal of the Bertrand Russell Archives*, N.S. 12(2), 197–207.

Rorty R. (1980) *Philosophy and the Mirror of Nature* (Oxford: Blackwell).

Sainsbury R.M. (1979) *Russell* (Arguments of the Philosophers Series) (London: Routledge & Kegan Paul).

Salmon M.H. (1974) "On Russell's 'Brief But Notorious Flirtation With Phenomenalism'", *Russell: The Journal of The Bertrand Russell Archives*, 16: 13–20.

Savage C.W. and Anderson C.A. (eds.) (1989) *Rereading Russell: Essays in Bertrand Russell's Metaphysics and Epistemology* (Minnesota Studies in the Philosophy of Science Vol. XII) (Minneapolis: University of Minnesota Press).

Schilpp P.A. (ed.) (1944) *The Philosophy of Bertrand Russell* (Evanston, Ill.: Northwestern University Press).

Scruton R. (1982) *Kant* (Oxford: Oxford University Press).

Slater J.G. (1976) "The Political Philosophy of Bertrand Russell" in Thomas and Blackwell (eds.), 135–54.

Smart J.J.C. (1963) *Philosophy and Scientific Realism* (London: Routledge & Kegan Paul).

Smart J.J.C. (1977) "Cosmic Coincidence", *Proceedings of the Russellian Society*, (University of Sydney), 2, 23–30.

Stroll A. (1967) "Identity" in P. Edwards (ed.), *The Encyclopedia of Philosophy*, Volume 4 (New York: Collier Macmillan), 121–24.

Thomas J.E. and Blackwell K. (eds.) (1976) *Russell in Review* (Toronto: Samuel Stevens, Hakkert & Co).

Tully R. (1988) "Russell's Neutral Monism", *Russell: The Journal of the Bertrand Russell Archives*, N.S. 8(1–2), 209–24.

Torretti R. (1978) *Philosophy of Geometry from Riemann to Poincaré* (Dordrecht: Reidel).

Urmson J.O. (1969) "Review of D. Pears, *Bertrand Russell and the British Tradition in Philosophy*", *Philosophical Review*, 18(4).

Urquhart A. (1988) "Russell's Zigzag Path to the Ramified Theory of Types", *Russell: The Journal of the Bertrand Russell Archives*, N.S. 8(1–2), 82–91.
Vuillemin J. (1972) "Platonism in Russell's Early Philosophy and the Principle of Abstraction" in Pears (ed.), 305–24.
Warnock G. (1969) *English Philosophy Since 1900* (Oxford: Oxford University Press).
Weitz M. (1944) "Analysis and the Unity of Russell's Philosophy" in Schilpp (ed.), 55–122.
Whitehead A.N. (1970) *The Aims of Education and Other Essays* (London: E. Benn; 1st edn., 1932).
Wiener N. (1964) *Ex-prodigy: My Childhood and Youth* (Cambridge, Mass.: M.I.T. Press).
Wilkerson T. (1976) *Kant's Critique of Pure Reason* (Oxford: Oxford University Press).
Winslade W.J. (1970) "Russell's Theory of Relations" in Klemke (ed.), 81–101.
Wood A. (1957) *Bertrand Russell: The Passionate Sceptic* (London: Allen & Unwin).

Index

Nijhoff International Philosophy Series

1. N. Rotenstreich: *Philosophy, History and Politics.* Studies in Contemporary English Philosophy of History. 1976 ISBN Pb 90-247-1743-4
2. J.T.J. Srzednicki: *Elements of Social and Political Philosophy.* 1976 ISBN Pb 90-247-1744-2
3. W. Tatarkiewicz: *Analysis of Happiness.* Translated from Polish by E. Rothert and D. Zieliński. 1976 ISBN 90-247-1807-4
4. K. Twardowski: *On the Content and Object of Presentations.* A Psychological Investigation. Translated from Polish with an Introduction by R. Grossmann. 1977 ISBN Pb 90-247-1926-7
5. W. Tatarkiewicz: *A History of Six Ideas.* An Essay in Aesthetics. Translated from Polish a.o. by C. Kasparek. 1980 ISBN 90-247-2233-0
6. H.W. Noonan: *Objects and Identity.* An Examination of the Relative Identity Thesis and Its Consequences. 1980 ISBN 90-247-2292-6
7. L. Crocker: *Positive Liberty.* An Essay in Normative Political Philosophy. 1980 ISBN 90-247-2291-8
8. F. Brentano: *The Theory of Categories.* Translated from German by R.M. Chisholm and N. Guterman. 1981 ISBN 90-247-2302-7
9. W. Marciszewski (ed.): *Dictionary of Logic* as Applied in the Study of Language. Concepts - Methods - Theories. 1981 ISBN 90-247-2123-7
10. I. Ruzsa: *Modal Logic with Descriptions.* 1981 ISBN 90-247-2473-2
11. P. Hoffman: *The Anatomy of Idealism.* Passivity and Activity in Kant, Hegel and Marx. 1982 ISBN 90-247-2708-1
12. M. S. Gram: *Direct Realism.* A Study of Perception. 1983 ISBN 90-247-2870-3
13. J.T.J. Srzednicki, V.F. Rickey and J. Czelakowski (eds.): *Leśniewski's Systems.* Ontology and Mereology. 1984 ISBN 90-247-2879-7
    For 'The Leśniewski Collection' *see also Volumes 24 and 44.*
14. J. W. Smith: *Reductionism and Cultural Being.* A Philosophical Critique of Sociobiological Reductionism and Physicalist Scientific Unificationism. 1984 ISBN 90-247-2884-3
15. C. Zumbach: *The Transcendent Science.* Kant's Conception of Biological Methodology. 1984 ISBN 90-247-2904-1
16. M. A. Notturno: *Objectivity, Rationality and the Third Realm: Justification and the Grounds of Psychologism.* A Study of Frege and Popper. 1985 ISBN 90-247-2956-4

Volumes 1–8 previously published under the Series Title: Melbourne International Philosophy Series.

Nijhoff International Philosophy Series

17. İ. Dilman (ed.): *Philosophy and Life.* Essays on John Wisdom. 1984 ISBN 90-247-2996-3
18. J. J. Russell: *Analysis and Dialectic.* Studies in the Logic of Foundation Problems. 1984 ISBN 90-247-2990-4
19. G. Currie and A. Musgrave (eds.): *Popper and the Human Sciences.* 1985 ISBN Hb 90-247-2998-X; Pb 90-247-3141-0
20. C. D. Broad: *Ethics.* Lectures given at Cambridge during the Period 1933–34 to 1952–53. Edited by C. Lewy. 1985 ISBN 90-247-3088-0
21. D.A.J. Seargent: *Plurality and Continuity.* An Essay in G.F. Stout's Theory of Universals. 1985 ISBN 90-247-3185-2
22. J.E. Atwell: *Ends and Principles in Kant's Moral Thought.* 1986 ISBN 90-247-3167-4
23. J. Agassi, and I.C. Jarvie (eds.): *Rationality: The Critical View.* 1987 ISBN Hb 90-247-3275-1; Pb 90-247-3455-X
24. J.T.J. Srzednicki and Z. Stachniak: *S. Leśniewski's Lecture Notes in Logic.* 1988 ISBN 90-247-3416-9
    For 'The Leśniewski's Collection' *see also Volume 13 and 44.*
25. B.M. Taylor (ed.): *Michael Dummett.* Contributions to Philosophy. 1987 ISBN 90-247-3463-0
26. A. Z. Bar-On: *The Categories and the Principle of Coherence.* Whitehead's Theory of Categories in Historical Perspective. 1987 ISBN 90-247-3478-9
27. B. Dziemidok and P. McCormick (eds.): *On the Aesthetics of Roman Ingarden.* Interpretations and Assessments. 1989 ISBN 0-7923-0071-8
28. J.T.J. Srzednicki (ed.): *Stephan Körner – Philosophical Analysis and Reconstruction.* Contributions to Philosophy. 1987 ISBN 90-247-3543-2
29. F. Brentano: *On the Existence of God.* Lectures given at the Universities of Würzburg and Vienna (1868–1891). Edited and translated from German by Susan F. Krantz. 1987 ISBN 90-247-3538-6
30. Z. Augustynek: *Time. Past, Present, Future.* Essays dedicated to Henryk Mehlberg. 1991 ISBN 0-7923-0270-2
31. T. Pawlowski: *Aesthetic Values.* 1989 ISBN 0-7923-0418-7
32. M. Ruse (ed.): *What the Philosophy of Biology Is.* Essays dedicated to David Hull. 1989 ISBN 90-247-3778-8
33. J. Young: *Willing and Unwilling.* A Study in the Philosophy of Arthur Schopenhauer. 1987 ISBN 90-247-3556-4

Nijhoff International Philosophy Series

34. T. Z. Lavine and V. Tejera (eds.): *History and Anti-History in Philosophy.* 1989 ISBN 0-7923-0455-1
35. R.L. Epstein: *The Semantic Foundations of Logic.* Volume 1: Propositional Logics. With the assistance and collaboration of W.A. Carnielli, I.M.L. D'Ottaviano, S. Krajewski and R.D. Maddux. 1990 ISBN 0-7923-0622-8
36. A. Pavković (ed.): *Contemporary Yugoslav Philosophy.* The Analytic Approach. 1988 ISBN 90-247-3776-1
37. A. Winterbourne: *The Ideal and the Real.* An Outline of Kant's Theory of Space, Time and Mathematical Construction. 1988 ISBN 90-247-3774-5
38. K. Szaniawski (ed.): *The Vienna Circle and the Lvov-Warsaw School.* 1989 ISBN 90-247-3798-2
39. G. Priest: *In Contradiction.* A Study of the Transconsistent. 1987 ISBN 90-247-3630-7
40. J. Woleński (ed.): *Kotarbiński: Logic, Semantics and Ontology.* 1990 ISBN 0-7923-0865-4
41. P. Geach (ed.): *Logic and Ethics.* 1991 ISBN 0-7923-1044-6
42. U. Wybraniec-Skardowska: *Theory of Language Syntax.* Categorial Approach. 1991. ISBN 0-7923-1142-6
43. J. Bransen: *The Antinomy of Thought.* Maimonian Skepticism and the Relation between Thoughts and Objects. 1991 ISBN 0-7923-1383-6
44. J.T.J. Srzednicki, S.J. Surma, D. Barnett and V. F. Rickey: *S. Leśniewski's Collected Works.* 1992 ISBN 0-7923-1512-X
    For 'The Leśniewski's Collection' *see also Volumes 13 and 24.*
45. P. Simons: *Philosophy and Logic in Central Europe from Bolzano to Tarski.* Selected Essays. 1992 ISBN 0-7923-1621-5
46. J. T. J. Srzednicki and D. Wood (eds.): *Essays on Philosophy in Australia.* 1992 ISBN 0-7923-1695-9
47. B. Dziemidok: *The Comical.* A Philosophical Analysis. 1993 ISBN 0-7923-2103-0
48. S. Read (ed.): *Sophisms in Medieval Logic and Grammar.* 1993 ISBN 0-7923-2196-0
49. K.J. Perszyk: *Nonexistent Objects.* Meinong and Contemporary Philosophy. 1993 ISBN 0-7923-2461-7
50. P.J. Hager: *Continuity and Change in the Development of Russell's Philosophy.* 1994 ISBN 0-7923-2688-1

Nijhoff International Philosophy Series

51. D.F.B. Tucker: *Essay on Liberalism.* Looking Left and Right. 1994
ISBN 0-7923-2705-5

Further information about our publications on *Philosophy* are available on request.

Kluwer Academic Publishers – Dordrecht / Boston / London

GPSR Compliance

The European Union's (EU) General Product Safety Regulation (GPSR) is a set of rules that requires consumer products to be safe and our obligations to ensure this.

If you have any concerns about our products, you can contact us on ProductSafety@springernature.com

In case Publisher is established outside the EU, the EU authorized representative is:

Springer Nature Customer Service Center GmbH
Europaplatz 3
69115 Heidelberg, Germany

Batch number: 10823959

Printed by Printforce, the Netherlands